The
Self Delusion

Also by Gregory Berns

What It's Like to Be a Dog

How Dogs Love Us

Iconoclast

Satisfaction

THE
SELF DELUSION

The New Neuroscience of How We Invent—
and Reinvent—Our Identities

Gregory Berns

BASIC BOOKS
New York

Basic Books
Hachette Book Group
1290 Avenue of the Americas, New York, NY 10104
www.basicbooks.com

Printed in the United States of America
First Edition: October 2022

Published by Basic Books, an imprint of Perseus Books, LLC, a subsidiary of Hachette Book Group, Inc. The Basic Books name and logo is a trademark of the Hachette Book Group.

The Hachette Speakers Bureau provides a wide range of authors for speaking events. To find out more, go to www.hachettespeakersbureau.com or call (866) 376-6591.

The publisher is not responsible for websites (or their content) that are not owned by the publisher.

Print book interior design by Linda Mark.

Library of Congress Cataloging-in-Publication Data
Names: Berns, Gregory, author.
Title: The self delusion: the new neuroscience of how we invent—and
    reinvent—our identities / Gregory Berns.
Description: First edition. | New York, NY: Basic Books, [2022] | Includes
    bibliographical references and index.
Identifiers: LCCN 2022019413 | ISBN 9781541602298 (hardcover) |
    ISBN 9781541602304 (ebook)
Subjects: LCSH: Identity (Psychology) | Self-perception. | Self-deception. |
    Memory.
Classification: LCC BF697 .B4595 2022 | DDC 155.2—dc23/eng/20220715
LC record available at https://lccn.loc.gov/2022019413

ISBNs: 9781541602298 (hardcover), 9781541602304 (ebook)

LSC-C

Printing 1, 2022

For Dad

A Note to the Reader

The work you are about to consume is an artifact. The author, as such, no longer exists.

Contents

PART III: **THE FUTURE OF YOU**

The Self Delusion

W HO ARE YOU? Asked this question, you'd probably respond with a name. But that's just a label—what others call you. At the core of this seemingly simple question lies a more complex one: What makes you you? Like asking a computer to think about itself, answering requires some computational contortions. Except that in this case, the computer is your brain. And like with Schrödinger's cat, the very act of observing our thinking has a defining effect on what we think.

Most people hear the "Who" in the question as singular, when in fact everyone has at least three versions of themselves. The first, situated in the present, is the one you're most accustomed to thinking of as "you." This you is reading this sentence. This you is acutely aware of your sensations. This you is conscious of the chair you're sitting in and how its cushions support your body, compressing and expanding with every shift in your weight. This you is

1

aware of the noises rattling around the room, or, if you're listening to an audiobook, the timbre of the narrator's voice.

But the present-you is an illusion. Present-yous just don't last long. How could they? Time marches forward, and in the milliseconds it takes to process the sensations like sitting in a chair, that you has already slipped into the past. Stub your toe, and it takes a couple of seconds before your brain even registers the pain.

Even when we think we're living in the present—as so many self-help gurus preach—we're really stuck in the past. This is the second version of you. Indeed, when someone asks who you are, the past-self is the one we automatically conjure up. It's the version of yourself whose lineage spreads back in time like the roots of a tree, providing the foundation of your identity. And though you may be tempted to define who you are with conventional labels like your job or what role you play in your family, the question of self-identity reaches deeper than any words you can use to describe yourself.

Although past-you stretches back to your earliest memories, it is not like a film recording. For one thing, memories can't be played in reverse. We can skip around and play bite-size clips, but these are merely the highs and the lows. For most of us, the whole of our past-selves—messy, complex, contradictory—has been reduced to a highlight reel. We make sense out of these fragments, imparting a narrative structure that seems to lead without interruption to the present-self.

Which takes us to the third, future, self. Although future-you is, by its nature, fuzzy, its function is both pragmatic and aspirational. When we think we're in the present, our brains are not only processing events that have already happened but also forming predictions about the immediate future. Every time you move, whether you're getting up off the sofa to get something in the fridge or you're crossing a busy street, substantial planning goes into the effort. The brain is, in fact, an engine of prediction, helping us stay one step ahead of a constantly changing world. This is one reason it's so hard to live in the present. Our brains evolved to predict the future.

Normally, past-, present-, and future-yous combine seamlessly into a unified existence. This, too, is an illusion, but it is a generally useful one because most of us don't change very much from day to day. Yesterday's you is likely very similar to who you are today and who you will be tomorrow. It's only over long spans of time that you begin to appreciate the differences among the three. When we look at pictures of ourselves, especially those from many years ago, the effect can be jarring. It often feels like we are looking at different people. That's because we are. The changes that occur from childhood to adulthood are so profound that at a physical—indeed, at a cellular—level, past-you and present-you are quite different beings.

So, who are you?

The answer is: whoever you *think* you are.

As we'll discover, the question of self-identity boils down to one of self-perception, namely, how a person thinks about themself. Take gender, for instance, which not so long ago was thought to be a fixed, objective feature of someone's identity. The argument went, you can look in the mirror and see for yourself the shape of your body, the genitals you were born with (or not). But we now understand that what you do with that information, your own sense of gender identity, whether congruent or not with the physical expression of your sex, is a matter of perception—and that is a process that occurs in your brain.

The neuroscience of perception has exploded in recent years. Advances in brain imaging have revealed that perception is an amalgam of physical reality filtered through the lens of past experience. It is an imperfect process because our senses give a limited view of the world. At every moment, the brain has to make its best guess as to the source of its sensory inputs. What objects are emitting the photons that hit your retinas? What animal or machine is making those sound waves vibrating your eardrums? Oftentimes the answer is indeterminate, and your brain has to rely on past experience to interpret what it is seeing or hearing.

The brain, though, is a computer with limited storage capacity. It compresses our memories into something like a low-resolution video, where reconstruction is prone to error and reinterpretation. Who you think you are, then, is shaped heavily by your memories of past experiences, flawed as they may be.

Which brings us back to the self-reflective computer: the puzzle of how we construct our identities boils down to how the brain thinks about itself.

AT THIS POINT, YOU MAY BE WONDERING ABOUT WHO I am—or rather, who I think I am—to insist that your sense of self is an amalgam of possibly buggy memories. Although I trained as a physician and psychiatrist, I work in computational neuroscience, which is just a fancy way of saying that I study how the brain computes things. Although the brain doesn't work like a man-made computer, it is a computer nevertheless. Current estimates place about eighty billion neurons in the human brain, and each one of these cells has the capacity to receive inputs from thousands of other neurons, perform some function on that information, and send its result to thousands of other neurons. Everything that we experience is a computation. The transformation of photons hitting the retina into the perception we have of gazing into the eyes of our loved one is a computation. The symphony of coordination between motor neurons and muscles that enables a professional golfer to hit a golf ball three hundred yards into the center of a fairway is a computation that results in the precise transfer of energy from body to club to ball. Even the feelings that we call emotions—like love, hate, shame, and joy—are computations that arise in the brain.

Although some scientists argue that the nature of these computations couldn't possibly explain the subjective experiences associated with them, let alone how they combine to create a personal identity, I hold a more optimistic view. Our knowledge of how the brain works is accelerating at a remarkable rate. I believe that we will soon understand how the symphony of neural activity gives rise to not only the emotions we feel but also our very identities.

Two factors are driving this revolution. First, new technologies like brain imaging have given us an unprecedented ability to observe the computations that brains perform. Scientists have begun to see how the computations relate to observable behaviors and, in the case of humans, interior experiences. Second, the incredible advances in artificial intelligence—AI—have been greedily incorporated into neuroscience. Neuroscience has been transformed from a science in which the goal was to locate specific functions in the brain to one in which machine learning is everything. Because of the complexity of the brain, neuroscience experiments generate enormous amounts of data—too much for any individual to understand. But it is a simple task to feed AI algorithms vast stores of neuroscience data and, using machine learning, find patterns that are otherwise invisible.

In this vein, self-identity—who we think we are—is a set of computations that our brains perform, with the distinguishing feature that the computations are directed at the brain itself. (I am sidestepping, for the moment, the question of whether "you" resides in your brain or some other part of your body. In the end, it doesn't matter because all sensations end up in the brain.) The challenge lies in decoding these types of computations. It is akin to what the philosopher David Chalmers called the hard problem of consciousness.[1]

It seems obvious that self-identity is a result of being conscious. When you're unconscious, you can't think about anything, let alone yourself. But when you're conscious, you are aware that you are an individual with your own sensations. This form of consciousness corresponds to present-you. Connecting this instantaneous form of consciousness to its past and future is what gives you a unique identity. To derive a theory of self-identity, we may not need to solve the hard problem of consciousness. We need to understand how the brain connects present-you to past- and future-you.

Humans have developed a unique cognitive skill to do just this: we tell stories. Storytelling is built into the human psyche, a process for which we have millions of years of evolution to thank. Our massive brains can do many things, like discovering DNA or sending people into outer space or creating transcendent works of art. Our species is able to accomplish such feats because of this most important skill, the ability to tell a story. No other animal can do this.

A story is a highly efficient way to represent a sequence of events, stripping down an event that happened in the real world into a compressed representation. You could say that a story is a curated version of reality. In the best stories, irrelevant information is discarded, leaving just the salient bits. To paraphrase Stephen King, a good story has "no detectable bullshit."[2] Indeed, good stories can be told with very few words. Although Ernest Hemingway probably didn't actually write it, he is said to have penned the six-word story: "For sale: baby shoes, never worn." The original version, attributed to William R. Kane, was even shorter: "Little shoes, never worn."[3] The amount of information packed into just four words is astonishing.

Good stories, though, do more than simply enumerate a sequence of events. Through narrative, they tell us what these events mean and why they happened. We use stories to understand the world around us and what is happening in our own lives. Meaningful stories let us store complex events of the world in our brains. We can use these stories to recall information tucked away over a lifetime. We can use these same stories to transmit knowledge to each other and to understand our place in the universe, or at least to construct one that makes life worth living. The alternative—that life is a series of random uncontrollable events—is too terrifying to ponder. H. Porter Abbott, a professor of English at the University of California, Santa Barbara, wrote that stories "are the principal way that our species organizes its understanding of time."[4]

And so it is with self-identity. Your identity is the story that connects past-, present-, and future-yous. That story is the cognitive construction that allows you to maintain the necessary illusion that you are, and always have been, the same person. How you tell this story, whether to yourself or other people, is the narrative that gives meaning to your life.

From my perspective as a neuroscientist, narratives amount to computations that our brains perform. Unlike facial recognition algorithms, which only require matching of images, narratives depend on the ordering of events into themes that make sense to us. Given the infinite number of events that might happen in the world, you might think that there are an infinite number of narratives, but this is not the case at all. Instead, as anyone who watches movies or reads novels can attest, our brains have evolved to slot these sequences into a relatively small number of narrative variations. And it's easy to get stuck in a familiar pattern.

When I was a practicing psychiatrist, every one of my patients found some aspect of their personal narrative distressing enough to seek help. Looking back, I realize that I lacked the proper tools to help many of these patients. Medications could help alleviate symptoms of anxiety and depression, but there was no drug that could course-correct personal narratives. This is beginning to change with the more widespread use of psychedelics, but it is not so simple to change course, even under the influence of drugs. A person's narrative is like a supertanker, massive and with correspondingly high momentum. Course corrections require advance planning and a lot of energy. Imagine a cargo-laden tanker cruising the ocean. The captain gets word that a hurricane is forming in his path. With enough warning, he can maneuver around the storm. Lacking sufficient time or fuel, he may have no choice but to plow through it, hoping for the best.

It's the same with narratives. Given enough time and energy, they don't have to play out in a predetermined way, but there are limitations on our ability to change them. As you might expect, long-standing narratives are more difficult to change than recently invented ones. More dogmatic and rule-bound people are less likely to shift narratives than those who are more cognitively flexible and open to new experiences. In fact, some of our cognitive intransigence may be a result of the first stories that we hear as children. If these early stories hew to moralistic plotlines, where the protagonist always does the right thing, this template can be hard to shake later in life. Also, people who are risk-averse, anxious, or depressed are also less likely to make course corrections in their personal narratives, and we now have evidence of biological changes that can explain why our brains become less plastic as we age. There are all sorts of reasons why it's hard to rewrite the story of one's identity, but the first step to change is learning how these narratives emerge.

THIS BOOK REVEALS HOW OUR BRAINS CONSTRUCT NARRATIVES for our lives and how this process constructs our self-identity. It's a complex journey, because a person's life encompasses not only everything that has happened to them but also everything they have heard or learned from other people. If it's in your brain, it's part of you, no matter how it got there. Add to that the ever-changing physical substrate for this information that we call the body, and you have a hard problem, indeed.

With that in mind, I've chosen to focus on five themes throughout this book that, though not comprehensive, will sketch the shape of an answer. Each of them appears repeatedly in different contexts.

EPISTEMOLOGY. Epistemology is the branch of philosophy that deals with knowledge and how we know what we know. While this may sound rather an abstract concern for nonepistemologists, the question of knowledge is central to who we think we are. For example, I have already stated that I am a physician and neuroscientist. That's a reasonable shorthand for who I think I am. But how do *you* know that's what I am? Because I told you so? This is a question to which I will return throughout this book.

COMPRESSION. The second theme follows from the first and deals with how the brain stores information. As I mentioned earlier, the brain is not a perfect recorder. Instead, it compresses memories using algorithms similar to those for streaming audio and video. The brain uses experiences laid down in childhood and adolescence as templates for future experiences. In this way, most experiential memories are stored as deviations from first experiences. Because your memories are foundational to who you think you are, understanding the compression/reconstitution process opens up an appreciation for how malleable identity is.

PREDICTION. Whereas compression is about the representation of past-you, prediction is about how the brain thinks about future-you. An explosion in neuroscientific information has demonstrated the myriad ways in which the brain acts as a prediction engine. Prediction is deeply wired into all animals' brains, allowing predators to catch their prey, or prey to escape being eaten.

In humans, prediction serves not only our basic survival functions but also when we are trying to outthink and outguess other humans. Prediction is the forward-looking face of identity. It forces us to think about our future-selves and how they will interact with other people. Prediction can be a source of both hope and anxiety.

DISSOCIATION. Dissociation carries negative connotations of psychiatric issues, but, in fact, everyone dissociates from time to time. Far from being a symptom of psychopathology, dissociation is a normal cognitive process. When thinking about past and future versions of ourselves, we have to dissociate from the present and put ourselves in the shoes of these other people we think we were or might be. The same process makes us feel like we are the main character in a novel or a movie. As frightening as it might sound, dissociation can be leveraged to alter one's perception of the past as well as to reimagine the future.

NARRATIVE. Here is the glue that binds everything together. We tell stories to ourselves and to others about who we think we are, and we listen to other people's stories in the hopes that they will shed light on the meaning of life. Throughout the book, I will return to the question of how we use narratives to organize information that flows into our brains. Narratives change the brain, regardless of whether they are self-told or encountered in the form of a novel or a movie. It's worth repeating that if it's in your brain, it is part of you. We'll discover how the provenances of these narratives floating about our heads get all mixed up, and how this soup of stories shapes our sense of self. And I hope to convince you that even if you think you know who you are, your narrative is still far from solidified. It's within your power to change it if that is what you want to do. The stories that you hear shape your narrative, and by changing the stories you consume, you inevitably shift yours, too. We will see how we can learn from past mistakes so that a future self will not regret what you do today.

By the end of our journey, you will see how these themes result in many of the traits we find desirable: maintaining a moral compass, living a good life without regrets, and steering clear of the false narratives that surround us. I hope you will have not only a better idea of who you think you are but also a sense of how to take control over the process and craft a narrative for the future-you.

PART I

THE SELF ACROSS TIME

CHAPTER 1

You Are a Simulation

E VERY YEAR I RUN A CLASS IN ADVANCED NEUROIMAGING techniques that teaches students how to design and conduct a functional magnetic resonance imaging (fMRI) experiment. It's a hands-on class, and in one unit they have to collect data on each other, taking turns going in the scanner. I am always bemused at the students' reactions to seeing their own brains. They display a curious mixture of excitement and disappointment. Of course, seeing your brain is thrilling, but exhilaration is inevitably followed by an odd disconnection from the contorted mass of tissue that makes you you. Like the moment when Toto pulls back the curtain to reveal the Wizard of Oz, you can't help but wonder: *Is that all?*

You literally cannot perceive anything about your brain. You can't see it from the inside, and you can't feel it. The brain interprets sensations coming from everywhere but itself. When you have a stomachache, you are acutely, painfully aware of things happening inside of you. You can poke and prod your belly, feeling your organs both from the inside and the outside. But your brain is inaccessible to touch and sensation. It feels nothing. As a result, you probably have more of a relationship with your spleen than with your brain.

Like the Wizard of Oz, the brain appears to construct a simulacrum—a low-fi simulation—not of itself but rather of every part of your body except the brain. And it is this construction that we perceive to be the self.

You, in other words, are a simulation.

Wait, you may be thinking, even if I can't feel my brain, it's still connected to my body, and my body is most definitely real. That is true. You can see your hands and feet. You feel the heat of the sun on your face. You taste chocolate on your tongue. You can move your body. Certainly, those are not simulations. If you knock a glass on the floor, it is a physical fact that it shatters. People around you will agree. We see each other and interact with the environment together, so there must be a physical reality to our existence.

But even physical reality poses problems. Two people looking at the same thing may disagree wildly about what they see. This perceptual divergence can be particularly acute when it comes to matters of subjective opinion, like art or music or food. And I'm not talking about whether one person likes something and another does not. Everyone has had the experience of comparing perceptions with other people and finding out that you saw or heard things very differently. Part of this has to do with embodiment. Even if two people witness the same set of events, because they don't inhabit the same body, and they don't see through the same set of eyes, each will have a slightly different vantage point. Most of the time, the differences amount to subtle distinctions. But not always. High-arousal states, like those that we experience in car crashes or during acts of violence, are well known to affect both perception and memory. This is the reason why eyewitness testimony can be so flawed and why it is essential to find corroborating witnesses to ascertain the truth, if there is such a thing.

I'll delve into these perceptual problems more deeply in later chapters, but for now let's focus on the question of you. More specifically, the question of how you perceive yourself and how that determines who you think you are.

You probably take your perception of yourself for granted. To take the simplest example, you know how you look. Unless you have impaired vision, all you have to do is stand in front of a mirror, and there you are. Although your image of yourself appears objectively real, your perception of how you look is in fact yours alone. Nobody else in the world sees you this way. That's because your perception is actually a mirror image of everyone else's. This is not a subtle distinction, as nobody is symmetric. You part your hair to one side or the other. When you speak or smile, your mouth turns up a bit more on one side. You may even, like me, have heterochromia, where your eyes are different colors. A photograph, on the other hand, captures the way that everyone else sees you. This is why, looking at a photo of yourself, it often feels like you are looking at a different person.

The misperception of self is even more pronounced when you hear an audio recording of your voice. Everyone thinks their recorded voice sounds nasal. I hate to break it to you, but I'm afraid that's how you really sound. Just like your image in the mirror, the voice you hear when you speak is your idiosyncratic perception. While the rest of the world hears your voice as carried by sound waves through the air, you hear additional sound from your mouth vibrating through the bones of your skull. It is a deep sonorous transmission because solid substances like bone conduct sound better than air. In our heads we all have a radio voice.

These misperceptions of self are immediately recognizable as such. And yet most of us continue to believe that the voice in our heads is what we actually sound like, the face in the mirror a faithful rendering of our appearance. Why should this be? Each of us constructs a picture of ourself that lives inside our head—and only in our head. Or, as I prefer to say, your brain is running a simulation of what you look like and how you sound. But a simulation is only as good as its inputs. Fed a steady diet of mirror images and bass-heavy bone conduction, your brain constructs a model of who it thinks it is and who the brain belongs to.

These examples illustrate distortions of the immediate type that shape your instantaneous perception of self. But, as I alluded to in the Introduction, the present-self is a chimera. The nerve impulses from the disparate parts of our bodies take different lengths of time to reach the brain. The notion that we are physically contiguous, then, is itself a construction of the mind.

Let's take another example. Take off your shoes and look at your big toe. For all practical purposes, light travels instantaneously, so the photons bouncing off your foot into your retina give you an immediate impression of your toe. (It takes another 10 to 20 milliseconds to reach your visual cortex, but we'll conveniently ignore that for the moment.) Now, go into the bathroom and run hot water over your toe. The nerve fibers that transmit heat and pain are the thinnest of peripheral nerves, which means that they conduct electricity slowly, about one meter per second. At that speed, it takes almost two seconds for your brain to know that your foot is burning up. But you can see the water hitting your toe instantaneously. How does the brain integrate these signals—one in the moment and the other two seconds in the past? The brain has learned over time that signals from your toe are delayed. As long as nothing surprising happens, the brain can form predictions about what your toe is going to feel two seconds in the future and make that jibe with what your eyes are telling you. Most of the time, this works fine. It's only when something unexpected happens, like stubbing your toe, that the system gets out of sync. You can actually see yourself mashing that toe and still have enough time to wince at how much it's going to hurt before the pain hits your brain. The present-self, temporally unified as it may seem, is spread out in time by a few seconds. It is only through cognitive processes and predictive neural processing that our bodies seem to exist synchronously.

The present-self, though, is not terribly important. Let's face it, our minds have difficulty focusing on the present, which itself is constantly flowing into the past, while our attention flits between what was and what will be. The present-self is but a two-second doorway between past-self and future-self.

Instead, let's take a closer look at our past-selves. We have two sources of knowledge about our past-selves: internal and external. The internal source derives from our memories. Although the nature of memory is still a mystery, scientists know a lot more about it today than they did just a decade ago. In broad strokes, there are different types of memory—some for facts, some for experiences, and others for muscle memory—and each of these memory streams is stored differently in the brain. Critically, you carry them around with you wherever you go.

These internal memories contrast with the external type, for which we can identify two important sources. First, there are recordings, including photographs, movies, audio, and written media like diaries and journals. Recordings are snapshots in time and have the advantage of being relatively immutable. As long as that second-grade school photograph of you remains intact, it can never change. The second external source derives from other people, specifically their memories. Unlike recordings, memories are constantly changing. When we gather with friends and family to reminisce about the past, it's often the case of the blind leading the blind. Who's to say what really happened at that Thanksgiving dinner ten, twenty, or thirty years ago? Your cousin is just as likely as you to misremember events from your childhood. But that doesn't stop us from filling in the holes in our memories with whatever source is available. Neurologists call this *confabulation*—filling in the lacunae of our memories with fictions.

By now, it should be clear that we have a problem. Our psyches are such that we feel, incontrovertibly, that our memories of our past-selves form a temporal continuum with the present-self. In other words, there is a seamless connection between the you reading this book and every instantiation of you back to your childhood. If it weren't so, then whose memories do you possess?

The physical evidence further complicates matters. A photograph of you from your childhood naturally looks very different from a more recent one. Neuroscientists have long known that the brain, like the rest of your body, keeps maturing until at least the mid-twenties, so it's clear we should question the reliability of our memories from childhood given that the physical substrate that contains them has also changed.

And yet we remain convinced of the accuracy of our memories, especially those that have acquired outsized significance: births, deaths, milestones, and traumatic events. Harvard psychologists Roger Brown and James Kulik called these pivotal moments flashbulb memories.[1] In the 1970s, they probed people's memories of the assassination of JFK and found that most people could recall what they were doing with an uncommonly high level of perceptual accuracy, as if the event had been burned into their brains like a flashbulb going off. Likewise 9/11.

As time went on, however, researchers began to scrutinize the accuracy of these memories. It turned out that flashbulb memories weren't all that accurate after all. In a longitudinal study of 9/11 memories, it was found that people forgot key details in just one year.[2] After that, accuracy declined more slowly, but even after ten years, people remained steadfast in what they believed they remembered. In other words, accuracy declined, but confidence remained high. Even worse, inaccurate memories were more likely to be repeated than corrected, further ingraining them into people's brains. It would seem that even our most vivid memories of key events in our lives contain a high degree of confabulation. Memories are like blind spots. The brain fills in what isn't there with a simulation.

If the past-self is riddled with holes, what about the future-self? The future is unwritten, but it is not completely unknowable. For example, you have a pretty good idea of what the weather will be like tomorrow. Barring catastrophe, you can probably predict with reasonable accuracy how your day will unfold. Our brains excel at these types of near-term predictions. On a moment-to-moment and day-to-day timescale, there is a high degree of continuity. Not much changes in our lives. So even though our memory systems are imperfect, they function sufficiently well that we can make plans to go to a concert or have a dinner out later in the week. It works because the world does not change much from day to day either.

Prediction becomes less reliable given longer timescales. What do you think you will be doing a year from now? Who do you think you will be in ten years? When I try to picture myself even twelve months hence, I get the unshakeable feeling of looking into a fog bank. Odds are good that I will be doing exactly the same thing that I am doing now—teaching, researching, and writing. I find this both reassuring and unsettling. Reassuring because the sense of stability is comforting. Unsettling because it augurs a finitude of life trajectory, like a projectile that cannot change course once it's launched.

The future is problematic because all we can do is rely on past experience to make predictions about what we will be. Whether we like it or not, our brains cannot help but take the past and project it into the future. The future-self is the purest form of simulation, but like the other examples I've described, the simulation is only as good as its inputs. And if the inputs are your memories, and if the memories themselves are flawed, then the future-self is also a fiction. It is a curious quirk of our memory systems that we tend not to remember our past predictions. Any decent AI system would have self-correcting mechanisms in place so that it could learn from its mistakes. Instead, humans suffer from confirmation bias. We remember when we were right and forget when we were wrong. So even though future-self is a fiction, it's not necessarily the case that the fiction—our predictions for what will happen—will become more accurate as time goes on.

So, where does that leave us? As the above examples reveal, it's evident that, at multiple levels, we are not the same person we once were. Our memories are flawed and our brains compensate by filling in fictions to connect all the slices of time, resulting in the illusion that we are a unified being on the journey of life. If this does not bother you, then you have achieved a rare state of equanimity and should put this book down. For everyone else, it's hard not to wonder how we can live with this simulation. As I'll show in the following chapters, new science has shown that we can in fact learn to steer the supertanker and gain some control over how our brains construct this delusion we call self. A glimmer of hope for a new narrative.

Before we can get to writing and rewriting personal narratives, we need to take a deep dive into the elements the brain uses to construct them. Memories, flawed as they may be, are the building blocks of narrative, but they still have to be slotted into a cohesive order that makes sense. In the next chapter, we'll see how the stories we hear as children provide the templates for arranging our memories. Because these early stories are the first narratives we encounter, they have a powerful influence that stays with us throughout life.

CHAPTER 2

Early Memories

F OR AS LONG AS I CAN REMEMBER, MY FAMILY HAD A TRA-
dition at Thanksgiving dinner. Around the time the pies were
brought out, my father would ask, "What is your first memory?"
When I was younger I dreaded it, but over the years it became a
running joke and we would all make up outrageous things to mess
with him. But if I'm honest about it, I could not state with any
certainty the first thing I remember. We had played this game so
many times, I was no longer sure whether an event came from my
own memory or from a story I had been told about my early life.

How strange it is that the period of life that developmental
psychologists claim is the most critical for the formation of identity is the very same stage for which our memories are the most
muddled. It's hard to contemplate the origin of these early memories, for they all get mashed together into the opening chapter of
our life story, repeated and passed down as family lore by parents
and elders.

In this chapter, I'll examine the lifelong legacy of the stories we hear as children. Parents use stories to tell their children about how the world works, offering information but also imparting expectations about what they want their children to be. Parents may not be conscious of why they tell the stories they do, or even the extent to which they tell stories to their children, but the impact of these stories is huge. Early childhood is the fountainhead of personal narrative because these early tales lay down templates to which we compare all other stories, especially the story of our life.

To understand the power of first stories, we must delve into the nature of memory itself and discover how we maintain knowledge through childhood and adolescence, while the brain itself is developing and changing at an incredible rate. As we'll find, the consistent consumption and telling of stories reinforce the idea that we're on a journey—one in which we are the hero.

BEFORE WE HEAD INTO THE SCIENCE, THOUGH, WE NEED TO take a brief detour into the nature of knowledge itself. How you know something to be true is a question to which we will return throughout this book, because this question is at the heart of who you think you are.

If someone states something as fact, and you lack firsthand knowledge of it yourself, it's natural to ask: *How do you know?* Bertrand Russell—a British philosopher, mathematician, and Nobel laureate—was one of the first to draw a line between two types of knowledge. There are things that you know how to do, like riding a bike, which Russell called knowledge by acquaintance. And then there is propositional knowledge, that 2 + 2 = 4, for example. But *how* do you know that? Because you learned it in grade school and everyone agrees that it is so. You also have acquaintance knowledge of basic math because you can count out jelly beans and see for yourself that if you add two jelly beans to two others, you get four. What about the fact that George Washington was the first president of the United States? None of us living today was alive then, so we have to accept such a claim as true. That is propositional knowledge, or what we would call simply: a belief.

A belief is an attitude that a person holds regarding something they think is true. It can be based on facts (although the veracity of facts is also subject to belief), or it might exist without proof, like a belief in the existence of God. In epistemology, the philosophical study of knowledge, the reasons people hold beliefs are called justifications. You might believe something because you saw it with your own eyes, you logically deduced it, or someone else told you so. Importantly, the justification of a belief can be conveyed only through a narrative, either to oneself or to others. The more closely we look at the nature of knowledge, the harder it becomes to separate knowledge from the stories we tell about what we know (or think we know). With this in mind, the first stories we heard as children become even more important. For not only do they form a template to which all other stories are compared, but they also stay with us throughout life, nudging our beliefs about things we accept as simple facts.

To answer my father's Thanksgiving question: my first memory is of being stuck at the bottom of a steep hill. I am three, maybe four, years old. Forest looms behind me, and I can smell the humus underfoot. I paw helplessly at the earth, unable to ascend the slope back to the playground. Yet I am not 100 percent sure of the veracity of this memory, because, strangely, I have no memory of how I got out of the forest. After so many Thanksgivings, the memory itself has been told and retold so often that it may have more in common with the feast's leftovers: a distorted version of the original, better in some ways and worse in others, but definitely not the same. True or not, this memory is a touchstone for my past-self and has become an essential part of the story I tell about who I am.

ONE OF THE MOST IMPORTANT, AND PROFOUND, DISCOV-eries about our brains is that there are different types of memories. In the 1980s, Larry Squire, a professor of psychology and neuro-science at UC San Diego, was studying stroke patients and how the location of their brain damage affected their memory. From this, he was able to identify at least two broadly different types of memory systems.[1] The first, which he called "nondeclarative," encompasses a wide variety of memories that require no language or labeling, such as simple Pavlovian responses and motor mem-ory, like remembering how to ride a bicycle or play a musical instrument. The second system, which he called "declarative," is responsible for two types of memories: knowledge of facts and knowledge of events. Fact knowledge—like the name of the first US president—is called semantic knowledge, while event knowledge—like childhood memories—is called episodic. Crit-ically, Squire found, the declarative and nondeclarative systems are based in different parts of the brain. Declarative memories rely on a part of the brain called the medial temporal lobe, which contains the hippocampus. Damage to the hippocampus results in anterograde amnesia—the inability to form new memories fol-lowing the injury. Nondeclarative memories, because they come in so many forms, rely on a wide variety of other brain systems.

The ramifications of multiple memory systems are both pro-found and disturbing. Each captures a different aspect of our per-sonal history. If our brains were accurate recording devices (and I don't believe they are, but it's a useful analogy here), then mul-tiple memory systems would be like tracks of a movie—different camera angles, dialogue, music, ambient sound, special effects. To-gether, they create a complete immersive experience. Separately, they give mere slices, and sometimes, they may appear to conflict.

Continuing the film analogy, the brain's task is to stitch all these memory traces together into a seamless narrative. As with a movie, the only way to do this is by editing. When memory research began to take off, most researchers focused on how and where the brain stored different types of memories, paying relatively little attention to how memories get sewn together. That is no longer the case. Scientists have become increasingly interested in the brain as editor.

Even as events unfold in real time, the brain lags in its interpretation. First, the memory has to be stored, which is called *encoding*. Some of this happens immediately, as evidenced by your ability to remember what happened a few moments ago. A memory in this state exists in a temporary holding buffer. It still has to be incorporated into the brain's long-term storage system, otherwise it will disappear forever. This second phase of memory storage is called *consolidation* and can take minutes to hours and, sometimes, days. Sleep plays a crucial role in consolidating the day's memories. Only after consolidation can a memory be *retrieved*.

The encoding/retrieval dichotomy is analogous to the Record and Play buttons on a camera, but recent research suggests that the processes aren't as separate as we once thought. Clearly, our brains don't function like video cameras recording every event in high spatial and temporal resolution. So how does the brain know what to encode? A leading theory, called transfer-appropriate processing, suggests that in any memorable situation, the hippocampus records the activity patterns of the cognitive systems that are active at the moment.[2] For example, memories of the *Challenger* explosion, or of 9/11, are largely visual. For those who saw these events unfold—generally on a television broadcast—their hippocampi were working overtime, binding the visual images together in a staccato sequence. Michael Rugg, a neuroscientist at the University of Texas, Dallas, has shown that when these images are called up from memory the hippocampus drives the visual system to replay them. The nature of episodic memory, then, is to repurpose the brain systems that were active at the time of the original experience by reactivating them in the same sequence. The process can also be triggered by experiencing something similar enough to the original event that the replay begins spontaneously—in other words, a flashback.

Given the critical role of the hippocampus in the formation of episodic memories, it's no surprise that the infant brain is not mature enough to store events. Nobody has a first-person memory of their birth. Childhood amnesia is a well-recognized phenomenon, and recent research is beginning to uncover its sources. At a biological level, the different parts of the brain develop at different rates. One way to track brain maturation is to measure the degree of myelinization in each region. Neurons are connected to each other via long projections called axons, and myelin is the waxy substance covering the axons and promoting electrical conduction in the nervous system. Infants are born with relatively unmyelinated brains. By our late teens, most of our brain systems reach their full myelinization, but the rate at which they do so varies greatly. Connections between the hippocampus and the structures that drive emotional processes seem to mature the earliest, probably by age five.[3] Similarly, the visual system reaches 90 percent of its adult connectivity by about the same time. The frontal lobes, which are associated with complex thought, are the last to reach their adult level of myelinization, usually in the early twenties.

It was once thought that childhood amnesia was dense, meaning we had literally no first-person memories from that early age, but work by psychologist Robyn Fivush at Emory has shown this isn't quite right. Fivush demonstrated that children as young as two and a half can remember events that happened six months earlier.[4] Two years old seems to be the hard limit before which nobody stores memories.[5] After that, the hippocampal system begins to come online, such that high-arousal events, like deaths, can be stored. By age four, it is almost fully functional, and the main period of childhood amnesia begins to come to an end.

We now know that childhood amnesia doesn't end all at once. Rather, it sputters out while the brain reaches its adult form. In the critical period of development from about age four to the early teens, there is a relative disconnect between memories and the systems engaged to replay them. On top of that, an adult brain recalling memories from this period is physically different from the brain that encoded them. It is like watching a movie recorded with VHS tape technology on a modern 4K-resolution display. This paradox raises the interesting possibility that unless you refresh your childhood memories from time to time, they might suffer the same fate as outdated digital media. (Floppy disks or zip drives, anyone?)

A larger paradox, though, derives from the fact that the present adult-self is physically different from the past child-self who encoded the memories in the first place. So much so, that we might reasonably ask whether those past-selves, especially far back in time, are the same person as our present-self. And if they are different, whose memories do we possess? With such a disparity, how does a person develop a cohesive narrative of who they think they are?

In the early 2000s, Elaine Reese, a psychologist at the University of Otago, New Zealand, began a landmark study on the origin of children's autobiographical memories.[6] Fifty children and their mothers began participating when the child was nineteen months old. The researchers visited the children's homes approximately every six months until the children were five years old, documenting changes in language and cognitive ability, with a particular focus on the emergence of the kids' memories. The mothers were able to provide confirmation that the child's memory was accurate. The initial findings were startling. Reese found that during this period, many children could recall events that happened before age three, and some could even remember things from before age two (although it was 50/50 as to whether they were accurate), contradicting the prevailing theory that nobody had memories from that period. After age two, more than 75 percent of a child's memories appeared to be accurate, but here's the kicker: by age five and a half, some of the earliest memories were already forgotten.

Years after the conclusion of the first phase of the study, Reese's team followed up with the children, now twelve years old. They began by asking the adolescent to describe their earliest memory and followed up with events that had been discussed during the first phase. As before, Reese found the adolescents' earliest memory dated to approximately two and a half years old, which was substantially younger than most adults' earliest memory. From this, Reese concluded that adolescents were still in the process of forgetting their earliest memories. This means that childhood amnesia could more accurately be described as a prolonged period of forgetting rather than a sudden event. But the kids varied in the degree to which they recalled memories. Reese writes, "The age of earliest memory was correlated with an adolescent's insight into life events and with their knowledge of family history."[7] In other words, the remembrance of childhood, both in quantity and in detail, is directly linked to social development—that is, learning to tell stories.

IT HAS BECOME INCREASINGLY CLEAR THAT CHILDHOOD IS marked by the disparate rates of development of parts of the nervous system. And while memory systems are coming online, the rest of the brain is changing. These maturation processes do not depend solely on laying down memories. In order to incorporate memories into a cohesive narrative, some things have to be forgotten. The dual processes of remembering and forgetting are powerfully influenced by the stories that a child hears and then later begins to tell themself.

The development of memory has received the lion's share of attention from researchers, but a few psychologists have dedicated their careers to the equally interesting study of how children tell stories. As I have suggested, the sense of self comes not only from memories but also from the narratives we construct to link memories together.

The period of childhood through adolescence is when memory, imagination, and forgetting coalesce into a mostly stable sense of self. Susan Engel, a psychologist at Williams College in Massachusetts, has written, "We are who we are by virtue of what we have experienced, but part of who we are is determined by what we imagine."[8] Through her work with children, Engel has been able to identify five phases of increasing sophistication in childhood storytelling.

First, toddlers learn they have an *extended-self*. As Reese's memory study showed, somewhere between ages two and three, kids learn that they have a past and that they can describe it. They know that their memories happened to them and that they existed in the past. As an adult, this may seem obvious, but connecting the past-self to the present-self requires cognitive hardware for mental time travel. This is the first step in constructing a concept of self that extends in time and space, a skill unique to humans in the animal kingdom.

The second phase follows closely, at about age three. Once a child understands themself as extending across time, they begin to incorporate contemporaneous events that happened to other people in their life—particularly family members. By hearing and telling stories to parents and extended family, the child begins to assimilate a version of the past that is based not only on first-hand experience but also on shared knowledge. As Reese's results showed, the degree to which families tell these stories has a direct impact on the number and density of memories from this period in a child's life.

In the third phase, children expand their extended-self beyond the family to include peers and friends. From age three to five, children share stories with each other, including events that happened to them personally as well as events that happened to other people. The social sharing of memories provides crucial feedback. Peers respond to events they find interesting, and children quickly learn what makes a good story: a premise that captures a listener's imagination and that also has an element of suspense, which makes them want to know what happened next. Initially, children don't differentiate between an event and their experience of it. A child in the third phase might tell a story of going to an amusement park as seen through their eyes, not yet capable of telling it from any other person's point of view. By the end of this phase, however, children learn that events don't simply happen. Usually they're the consequences of a person's actions.

The fourth phase is the most critical. From about age five through nine, children increase their repertoire of stories. They try them on like outfits, looking to parents and peers for feedback as to which ones fit the best. Children in this phase are often surprisingly good storytellers. Not necessarily because of their skill in constructing a plot, but because of their unselfconscious recollection of personal details. An example from Engel's work illustrates the power of a child's observations, even in a simple story:

> I had a car that I rode around the house. When my Mom and Dad had a fight I pushed them away from each other. Then my Dad moved to Albany. When I was a little kid every time I sat in the tub I was afraid I would go down the drain.[9]

ON THE SURFACE, THIS IS A SEQUENCE OF THREE EVENTS plus an introspection in an unrelated scene. But the fact that the child chose to string them together in this way makes them all connected. The power of the narrative comes from the construction itself. As Engel noted, this childlike simplicity is also the hallmark of some great writing.

Ernest Hemingway was famous for his simple narrative structure. In *The Sun Also Rises*, Hemingway tells the story of Jake Barnes, an American writer living in Paris after World War I. After a night of heavy drinking in which Jake and his friend Robert come to blows, Hemingway writes (as Jake):

> *I could not find the bathroom. After a while I found it. There was a deep stone tub. I turned on the taps and the water would not run. I sat down on the edge of the bath-tub. When I got up to go I found I had taken off my shoes. I hunted for them and found them and carried them down-stairs. I found my room and went inside and got undressed and got into bed.* [10]

THE SIMILARITIES TO THE CHILD'S STORY, IN BOTH STRUC-ture and topic, are striking. A drunken attempt to take a bath becomes unexpectedly poignant in its simplicity.

Narratives tell us why things happen, or, as in these examples, they give us insight into a person's state of mind. Hemingway's technique, like that of many children, demonstrates that good narratives do not require sophisticated language. It also suggests that the stories from our childhoods, both the ones we told as well as the ones we heard, stay with us throughout our lives. Their power stems not only from the content but also from the structure itself.

The narratives children encounter from the critical period between the ages of five and nine come from two sources: events they experience themselves, and stories they hear from other people. Every day brings with it new experiences to be integrated into a rapidly changing brain. The research by Reese, Engel, and others shows that storytelling is not just a side effect of the development process. It *is* the process. And the stories that are laid down from about age seven to nine may be some of the most enduring because that's when the brain—as well as the rest of the body—begins to approach its adult form.

The final phase in Engel's framework, from about age nine to the onset of puberty, marks the period of stabilization. As children settle into patterns that form the bases of their identities, their repertoire of stories decreases. Take Santa Claus. The absurd premise requires a suspension of all knowledge about the relative sizes of a man and a chimney flue. Until a certain age, children readily accept this possibility. When reality dawns on them, they may go along with the charade for a year or two, but then no more. Such a decrease in range of acceptable stories marks the beginning of the editing process. After that, children are unable to simply experience life without precondition. Events and memories are increasingly slotted into narrative forms they have already encountered. Although the loss of open-ended possibility feels limiting, it has to be this way. It's the only way to preserve a sense of self that is already unsettled before the hormonal onslaught that will shake it up like a snow globe.

These early stories, so critical for the formation of self, lay the foundation for the stories people tell throughout their lifetimes. Because these stories create templates for all stories that follow, they implicitly bias the perception of each new bit of information. The significance of incoming events may not be judged by the objective truth of the events but by how well they fit an ongoing narrative. And when an event doesn't fit the narrative? There are really only two options: change the narrative or discard the event. We already know that people don't like to change their stories. In the next chapter, we'll learn how the brain uses these narrative templates to process information, editing events as they happen, and ultimately constructs our sense of self.

CHAPTER 3

Compression

EVERY DOCTOR REMEMBERS THE FIRST PATIENT THEY LOST.
I was a first-year intern at the largest hospital in Pittsburgh.
I admitted an elderly patient with a swollen leg, not an obviously complicated case. He had history of blood clots in his leg, and an ultrasound confirmed a deep vein thrombosis. Per the standard treatment protocol, the second-year resident and I administered anticlotting medication. But what we didn't realize is that the patient had banged his knee prior to coming to the hospital and that the swelling was due to the injury, not the clot. Our treatment only made the situation worse. The bruise on his leg started bleeding into the surrounding tissues, choking off circulation so badly that he needed surgery to release the pressure. After the procedure, he bled into his brain and died.

I remember slumping into a chair, wishing I could go back in time. Of course, I couldn't. Now, this patient's memory lives on in my brain as a compressed version of what happened, a lesson about the hubris of medicine. This mnemonic compression continues to fascinate (and humble) me because it reveals what our memories, and ultimately our narratives, are really made of.

We cannot help but feel that our memories are accurate recordings of things that happened to us. But this is a function of filling in the gaps between memory slices, as I discussed in the last chapter. Having said that, the brain does behave in at least one way that is similar to video. The brain takes snapshots, kind of like the individual frames of a celluloid film. When these snapshots are called up from the memory banks, the brain acts as an editor, stitching the snapshots together to create what feels like a seamless narrative. It's here, during the editing process, where a narrative emerges and the past-selves are linked together.

The brain, though, is an imperfect editor. Our memories are, at best, compressed recordings of events. To reconstitute actual events in the mind's eye, the brain needs a template, some sort of storyboard on which to arrange its snapshots. In the last chapter, I suggested that the stories we hear as children form the bases of some of these templates. Now we'll dive into the particulars of how this happens neurologically. There are profound implications for the construction of self, which, we'll discover, is riddled with holes.

ALTHOUGH I USE COMPUTER ANALOGIES FREELY WHEN talking about the brain, computers and brains differ in one important aspect: the user. Engineers continue to evolve mechanical and digital technology so that computers can do more and more for us, but there is no mistaking who is in charge. Until computers become self-aware, they remain tools for humans. As I pointed out in the first chapter, our relationship to our brains is quite different. The brain is the computer that makes us who we think we are. It's impossible to separate out the user from the technology.

We're left with a mind-bending question: How does the brain think about itself?

I'm afraid that neuroscience hasn't figured this out. However, even if we can't directly answer the question, we can still explore around it, identifying ways in which the brain *doesn't* think about itself. For starters we can point to how the brain's construction places limits on what we can know about ourselves.

Consider the exquisite complexity of the bodies that contain our memories. There are roughly 10^{27} molecules in an average person. For all practical purposes, those molecules could be arranged in an infinite number of ways. Precious few of these arrangements are compatible with life, and fewer still result in sentient life. And of these possible arrangements, only a rare handful could create the unique human being that each of us believes we are. As my patient taught me, there are many more ways to be dead than alive. All the doctors in the world could not bring him back to life. When I gave him medication to break up the clot I thought he had, a series of chemical reactions was set in motion that would result in his atoms being turned to dust in a shockingly short period of time.

A person's death brings home the fragility of life, to be sure, but it also reminds us of the life they lived. What happens to the experiences contained in a person's brain? Are they lost in the entropy of their decaying flesh?

Not entirely.

Even if we can't reconstitute a person's molecules, we can capture and store low-fidelity facsimiles of that person. We take pictures and movies of each other, which are nothing more than arrangements of molecules in a physical medium that our brains recognize as an approximation of a person. Our synapses get rearranged, constituting the memories in our heads about them. We tell stories to each other, and if these stories are interesting enough

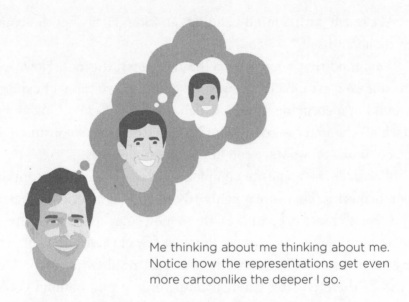

Me thinking about me thinking about me. Notice how the representations get even more cartoonlike the deeper I go.

to other people, they will keep the stories in their memories, too. Some of these memories are written down and preserved in other ways, too.

Even though I don't remember the name of the patient with the swollen leg, the end of his life story became embedded in my brain. He lives there with all the other ghosts. In this sense he is still very much alive—I can imagine this avatar speaking to me and chastising me for making unfounded assumptions about disease processes and treatment plans. And now, by virtue of me committing his story to the page, he exists in the heads of everyone reading this book.[1]

Hard as it may be to accept, our own personal narratives are in fact no different from these ghosts. I count myself among a growing number of scientists who believe that the construction of self-identity is not much better than the lo-fi representations of other people we hold in our heads. Computer scientists call this a recursion problem. Let's assume that the brain does all the work. It doesn't matter where the computations occur, just that they occur within the body. If your brain contains "you," then it must also contain "you" thinking about your brain containing "you." And so on.

If you find this hard to think about, don't worry. I get a vague sense of vertigo whenever I think too much about thinking. Perhaps our software has built-in protection from going too far down the rabbit hole. Or maybe evolution killed off any creatures who got stuck in this endless existential loop. In fact there is a simpler explanation: It's not possible for a computer to represent itself exactly. If it could, it would need a computer of equal size, and that would need an auxiliary computer, too. Actually, you would need a bigger computer—to contain the copy of the original plus the link to the copy. By physical necessity, the brain's representation of "self" must be a low-fidelity version, because that is all it *can* contain.

Because of the computational limitations of our brains, any knowledge we have, including the knowledge of our own narrative, is represented in a compressed, reduced format. Our conceptions of self and others are cartoon versions of the real things. These caricatures gloss over the nitty-gritty changes that occur from moment to moment, allowing us to maintain the illusion of continuity—that we are the same person we were yesterday. The brain, in other words, is designed to forget.

A rather disturbing conclusion emerges from this line of thought: Who you think you are—your notion of "self"—is a mere cartoon, just as your notions of other people are cartoon versions of them. But critically, these cartoons form the touchpoints of narrative that bind together our mental models of the world and our place in it.

THE NARRATIVE THAT BINDS OUR PAST-SELVES TOGETHER with the present-self has to be threaded through time in an order that makes sense to us. A story is just a sequence of events, but a narrative imparts meaning to the sequence, which requires certain assumptions about causality. We assume that if event A occurs before event B, A might have caused B, but never the other way around. Every narrative that we construct is predicated on the assumption of causality. Causality is the reason for constructing narratives in the first place. It lets us build models of how the world works and, in doing so, allows us to make predictions about the future. If B always follows A, then we can be pretty certain that if A appears, B will soon follow. But abstract symbols, like A and B, are hard to remember. It's much easier to recall stories about why things happen the way they do. This is how superstitions arise, like the taboo against walking under a ladder. It's very likely that someone, at some time, walked under a ladder that then fell on them, leading them to believe that the act of walking under it caused it to fall.

Is it possible that a falling ladder caused a person to walk under it? If you think about it for a moment, you can imagine a scenario in which that is exactly what happens. The ladder starts to fall, and you rush under it to prevent it from crashing through a window, but in doing so, you get hit by it instead. You can see how our notion of causality depends on one's perspective. Notions of causality also depend on initial conditions, like whether the ladder was stationary or moving. But we often don't know what the initial conditions are, so our conception of causality is necessarily based on what we see with our own eyes, or what someone tells us.

When something happens, our perception of it is not the same thing as the event itself. It's the difference between watching the Masters on television and watching it in person, versus being Tiger Woods winning for the fifth time. Anyone can tell the story, but everyone's version will be different. Usually, we can agree upon the order in which things happened. Tiger played seventy-two holes of golf in the specified sequence. The ball never left the fairway to fly back to the tee, nor did the ball ever jump out of the hole to connect with his putter. The arrow of time is the same for everyone. (That's a good thing. Imagine how difficult it would be if we moved backward and forward in time. The notion of a story would become meaningless because events would no longer happen in defined sequences.) However, there are circumstances in which our perception of the order of events can get turned around. When that happens, notions of causality become reversed. Imagine you were riding on the head of Tiger's driver. Rather than the driver hitting the ball, you might think the ball hit the driver! And if that were your point of view, you might construct a very different narrative for what happened: *I was minding my own business, when, all of a sudden, this ball smacked me in the face!* Like the falling ladder, perspective changes the interpretation of events, and perspective determines our notions of causality.

Many years ago I was involved in a minor fender bender in a parking lot. My version of what happened goes like this: I was getting ready to back out of a parking spot. The traffic lane was one-way only, so I looked behind me and to the right—the direction from which I expected traffic to be coming. With the coast clear, I started backing out. Imagine my surprise when I heard the sickening crunch of metal on metal. I stopped immediately, realizing that someone had entered the traffic lane from the wrong direction—in my blind spot on the left side. The other driver's version went something like this: I was driving along, minding my own business, and, for no reason, this car backed into my side door.

No one was hurt, and the impact was at such a low speed that the apparent damage was minimal, if not minimal in cost to repair. Our insurance companies battled it out for over a year only to resolve it at arbitration, where I was deemed at fault for failing to determine an unimpeded path to back up. If I had backed out a second sooner, the other driver would have collided with me, and the final narrative would have been very different.

We can see how slight changes in perspective can result in wildly different interpretations of events. How we tell a story—whether to ourselves or to other people—dictates our interpretation of a sequence of events. But if our interpretation of events is apparently so fluid, how can we agree on anything?

The answer hinges on the arrow of time. Events happen and, because we cannot go back in time and revisit them, we need an efficient way to keep track of as much as we can. But our brains are not limitless memory banks. We can't record everything that happens, and we can't replay what we've recorded at will. So we organize events into preexisting narrative forms. The arrow of time forces us to winnow out the space of narratives and separate them into types that conform with our observed order of the world. Think of these narrative forms as storyboards that contain the cartoon versions of ourselves and other people that we hold in our heads. As the research of Reese and Engel shows, these narrative forms begin to take shape in childhood and reach something close to their adult form by the onset of puberty.

That still leaves plenty of room for error. When I slotted my patient with the swollen leg into a narrative that I had learned in medical school, I failed to realize the importance of the banging of his leg. How could I not? Nobody had told me such a story. I had to learn that one through experience. Likewise, when I backed up my car, I was operating the vehicle according to a narrative in which other drivers always obey traffic directions. I suppose I had been fortunate to not have encountered wrong-way drivers up to that point, but now I know better.

These are not unreasonable narratives. If we had to consider the universe of possibilities at every juncture in time, we would be unable to leave the house. Narratives let us represent complex phenomena in highly compressed fashion. Here are a few narratives compressed into one phrase. Yet consider how much information they contain:

- It's a tale of rags to riches.
- What an incredible Cinderella story (cue Bill Murray in *Caddyshack*).
- He/she has Daddy issues (different narratives, depending on gender).

Stereotypes? Absolutely. But all narratives are stereotypical. They have to be, because our brains aren't video recorders. You don't even need particularly complicated psychological mechanisms to start building narratives. All you need is a brain with a little bit of memory. From there, you can construct mental models to try to make sense of the world.

IN THE WORLD OF COMPUTER SCIENCE, THESE STEREOTYPES are called *basis functions*. An XY graph is a simple basis function. It has two dimensions, and every location in the plane can be represented by its coordinates. A circle, for example, can be described by a basis function using x and y coordinates: $x^2 + y^2 = \text{radius}^2$. More complicated things—like the heart's electrical activity on an electrocardiogram—can also be represented by basis functions. One of the most important discoveries in the math of basis functions was made by French mathematician Jean-Baptiste Joseph Fourier. In the early 1800s, he discovered that any mathematical function could be represented by a combination of sine and cosine waves of different amplitudes and frequencies. The Fourier transform can take something like an ECG and turn it into a basis set of sine and cosine waves. A similar type of operation is at the core of the compression routine that forms JPEG images. Individual pixel values, which take up a lot of memory, are transformed into a collection of cosine waves. These basis functions require much less memory to store than the original.

The same type of operation is at work in the brain. There is not enough memory to store every episodic memory in its original form. Like a JPEG (or the movie equivalent—MPEG), the brain uses basis functions to store the compressed representations of memories, that is, the snapshots that constitute an episodic memory. Psychologists call these compressed representations *schemas*.[2] Because a schema is an abstract representation, it guides both the recall of old information and the encoding of new events. Neuroimaging studies have shown that a portion of the prefrontal cortex along the midline—the ventromedial prefrontal cortex, or vmPFC—uses schemas to influence the processing of sensory information coming into the brain.[3] How this happens has not been completely worked out, but recordings of electrical activity suggest that the vmPFC affects activity in sensory regions through relatively slow oscillations called theta waves (4–8 Hz).[4] These waves appear to help synchronize activity in the different sensory systems encoding a particular memory, like matching up the audio and video tracks of a movie. Once a schema is in place, it biases what a person sees and hears to fit with an existing template. This is why I saw my patient with a swollen leg as having a DVT instead of potential compartment syndrome. Schemas also bias what is encoded into memory. If something doesn't fit with an existing schema, then it may not be remembered at all, or it may be remembered in a way that best fits with an existing template.

Work by Olivier Jeunehomme and Arnaud D'Argembeau, psychologists at the University of Liège in Belgium, has revealed how the continuous stream of daily events is chopped up into pieces and slotted into schemas. They had students wear GoPro-like wearable cameras while walking around campus and performing various activities.[5] Afterward, the students had to recall what they did while the researchers recorded their narrative. The students then had to match elements of their recorded narrative with frames from their videos. What the researchers discovered was that, rather than being continuous, episodic memory was defined by event boundaries—when and where something changed. These boundaries behave like stops on a journey. By using the video frames as references, Jeunehomme and D'Argembeau were able to estimate the degree of memory compression by the number of remembered events per minute versus the actual number of events recorded by the cameras. Interestingly, they found that moving around or sitting still was compressed the most—by a factor of five—while other actions were not compressed at all. Their results demonstrate that our basis sets for schemas are organized around things that we do and places we go.

The advantage of schemas lies in their efficiency. Once a schema has been created, new events only have to be processed and stored as deviations from the template. Everyone remembers their first kiss. That created a schema by which all other kisses are referenced. Can you remember your second kiss? Probably not very well. That's because it was encoded as a deviation from the first one. You need the schema plus the deviation to reconstruct the event.

Even the memory of your first kiss has been corrupted by preexisting schemas laid down by childhood stories. Look how many tales center around the magic power of a kiss: *Snow White* (although in Grimm's original version, the prince carries Snow White home in her glass coffin and accidentally dislodges the slice of apple stuck in her throat, causing her to wake up), *The Princess and the Frog*, and more recently, the movie *Shrek*, where only "true love's kiss" will break the curse that has turned Fiona into an ogre (in fact it makes her stay in ogre shape forever). Although we all remember the magic power of the kiss in these stories, that's but a snapshot of the tale. The kiss is just a mnemonic tag that unravels these stories. And although you may not consciously think of these stories when you remember your first kiss, they are there, entangled with your experiential memory.

By early adulthood, your brain is riddled with schemas. They are idiosyncratic because they are based solely on things that happened to you and the stories other people told you. There is no way to purge them. Most of the time, this is a good thing because schemas help us make sense of the world. They provide an organizing framework for what would otherwise appear to be random events. These compressed representations make you "you," or, at least, who you think you are.

Although schemas form the basis of efficient memory storage, they raise a question about how much of your past-self is scaffolded by stories versus how much from past experience. As you might imagine, it is not straightforward to disentangle these two elements. It's much easier to demonstrate how these schemas affect the present-self. In the next chapter, we'll see how the stories that link past-selves also affect the immediate perceptions of the present-self.

CHAPTER 4

The Bayesian Brain

IT'S NOT A STRETCH TO BELIEVE THAT OUR MEMORIES CAN be inaccurate, muddled by other people's accounts and past experiences. More surprising is that even our instantaneous perceptions are fraught with inaccuracies. We see what we expect to see, including in ourselves.

The present-you is a sort of machine that converts things happening around it into internal representations. If you are reading this as printed text, photons are striking the page and bouncing into your retinas. If you are listening to this as an audiobook, sound waves are hitting your eardrums. These physical reactions are transformed into something meaningful in your brain. And what exists in your brain is not a recording of these events but a highly processed representation of them. It's through the process of perception that these external events are transformed into internal representations. If your life is like a movie, then these perceptual events are the individual frames, and how these frames are linked together determines the movie's plot, that is, its narrative. We begin by dissecting a single frame.

Perception is a psychological process to which we generally don't give much thought. It operates below the level of conscious awareness, and yet, perception is the very essence of how we interact with the external world. Perception encompasses the processes by which sensory organs convert external energy into neural signals and how the brain then reconstructs the sources of those signals.

Take your vision, for example. Look around. Find an object—a piece of fruit or maybe a favorite coffee mug. Without touching it, examine it closely, taking note of its shape, its color, its texture. When you are sure that you have seen everything there is to see, turn the object upside down. Look at it again. Do you notice things that you didn't see before?

The point of this exercise is to illustrate that when you look at something, especially if it's familiar, you don't see many of the details that make the thing unique. An apple is an apple, right? Categorically, that is true, but each individual apple is also unique, distinguishable by its specific coloration, its shape, its size. In other words, our perception of a thing is shaped by our prior beliefs about what kind of thing we are looking at.

Another example of how prior beliefs affect perception is the optical illusion known as the Kanizsa triangle. When you first look at the image, the dominant perception is that of a white triangle occluding three circles and an inverted triangle. You may even believe that you can see the actual outline of the top triangle. But this is an illusion created by your brain. You can temporarily disrupt this illusion by focusing on the circles. Instead of considering them as circles, though, think of them as pie charts or, if you are of a certain age, Pac-Men. It isn't easy. The white triangle always reasserts itself as the dominant perception. An even more powerful illusion is Idesawa's sphere. Here, the illusion leaps off the

Kanizsa's triangle and Idesawa's sphere.

page, creating the impression of a three-dimensional, spiky sphere. There is no effective way to make the illusion disappear.

A common explanation for these illusions comes from the Gestalt psychologists of the early twentieth century. *Gestalt* means "form" or "pattern" in German, and the psychologists of this school believed that human perception is primarily a top-down process of imposing form on the whole of the image. This was in contrast to the preexisting view—call it bottom-up—that perception is a forward process of assembling low-level elements of vision, such as lines, shapes, and colors, into an aggregate concept that exists only in the mind's eye.

One problem with the Gestalt theory is that you would need to know about triangles and spheres in order to reconstruct them from the sparse visual input of the illusions. In other words, you can't see what you don't know. The bottom-up theory suffers from the same shortcoming. How does your brain build a representation of things—whether a triangle, a sphere, or an apple—if you didn't already know what they were? We don't perceive the world as visual primitives. We perceive the world as scenes filled with people and things.

Hermann von Helmholtz, a German physicist and physiologist, recognized in the nineteenth century that perception was fundamentally a statistical problem. Focusing on vision, Helmholtz realized that the information from the eyes did not uniquely determine what a person was looking at. The brain had to solve the inverse problem: Given the stream of photons hitting the retina, what is the most likely source?

For instance, imagine you see a person standing at the end of a hallway. They look rather tall. There are two distinct conclusions you can make about the person. The first is that the person is actually tall. The second is that they are of normal height, but they are close to you. How do you decide? In the real world, plenty of other visual cues usually allow you to assess the distance to the person. But ultimately, you have to make an educated guess based on the relative likelihood of encountering a tall person or a person of normal height.

Thomas Bayes, an eighteenth-century English statistician, discovered the mathematical rule that shows how this happens. Bayes' rule says that when we obtain new information about an event—like seeing a person in a hallway—we should update our prior expectations on the basis of the information. Blackjack provides a good example. With a fresh deck of cards, the odds of hitting 21 with two cards is about 1 in 20. But suppose your first card is an ace. Now there are sixteen cards worth 10 remaining, and so the odds of drawing one of them is 16 out of 51 remaining cards, or about 1 in 3. That is Bayes' rule—you use the new information (the ace you've drawn) to update the probability of a specific outcome.

In the early 2000s, neuroscientists started taking probabilistic theories of perception seriously by looking for evidence that this was, in fact, how the brain constructs perceptions. Since then, the evidence has continued to grow, and the idea of the Bayesian brain has become the most popular theory of perception. The core idea is that the brain is constantly guessing the posterior probability, which is the probability that you are looking at a tall person given the particular set of images hitting your retinas (the probability is called "posterior" because it occurs after the receipt of new information, in contrast to the original, or "prior," probability).

It sounds complicated, but these are the types of mathematical operations that neurons are well suited for. Bayesian inference doesn't require conscious knowledge of the specific probabilities. These could be encoded from past experience and exist well below the level of conscious awareness. Visual priors are stimuli that you have learned are consistently associated with physical realities of the world. One common prior is the fact that light comes from above, and we use this knowledge to determine whether a curved surface is convex or concave. Another common prior is that when a shadow moves, it is because the object is moving, not the light source. There is even evidence that our visual systems are tuned to the spatial frequencies that are present in natural scenes.[1] What that means is, when presented with potentially ambiguous visual information, our brains are more likely to interpret it in a way that is most consistent with what occurs in nature.

A classic example of the use of visual priors occurs in the crater illusion. When looking at the left figure below, most people see a crater—or concave indentation—while the right figure appears convex, like a mound surrounded by a moat. If you flip the image upside down, the crater and mound reverse. The best explanation for the illusion is that our brains assume that the light source is always coming from above.

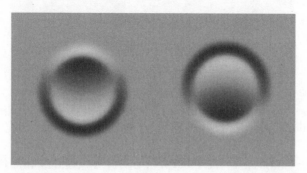

Shaded figures appear concave (left) and convex (right). The images are identical except for being rotated 180 degrees. The illusion comes from the brain's assumption that light always comes from above.

But notice what happens when you rotate the figures by 90 degrees. The percept—the result of perception—is less stable. Both figures may appear concave, depending on where you focus your attention, or one may appear convex, but this, too, can shift back and forth. It all depends on where your brain assumes the light is coming from.

These illusions reveal two important aspects of Bayesian perception. First, when the prior is strong, as in the assumption that light comes from above, the prior will dominate the percept. When the prior is weak, as when the light could come from either the left or the right, the prior will have relatively less impact on the final percept. Second, even when the percepts flip back and forth, you can experience only one at a time. It is either convex or concave. There is no in-between. This means that even though the brain can hold multiple interpretations of a sensory input, only one rises to conscious awareness. In probability theory, this is called "winner take all," meaning that the most likely percept, even if it is only slightly more probable than another, is the one that we see.

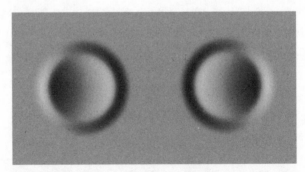

When the figures are rotated 90 degrees, the concavity or convexity becomes unstable and can switch back and forth.

These visual tricks have been known for decades, and the Bayesian explanation offers a sort of folk theory for perception. Yet only in recent years has neuroscience taken the idea of the Bayesian brain seriously. The critical question has been: What evidence is there that the brain actually uses Bayesian statistics to form percepts? It's a hard question to answer because we don't really know how the brain encodes prior knowledge. But even without directly addressing knowledge representation, figuring out how the brain encodes uncertainty offers us some clues toward an answer. In different experiments, researchers varied the predictability of simple visual images presented to animals while recording activity in neurons in their visual cortex. Aspects like contrast (more contrast makes images more certain) or the addition of noise (making images less certain) caused measurable changes in visual neurons. The more uncertain the image, the more variable the responses of the neurons.[2] It appears that the neurons themselves cycle through various interpretations of the information they receive. The more ambiguous the input, the more possibilities they consider.

One criticism of the Bayesian model points out the apparent impossibility of storing probabilities in the brain. A perfect Bayesian brain would not only keep a record of all the things that happened to an individual but also associate probabilities with events. To do that, it would have to keep track of how many times something happened relative to all the other things that had, or could have, occurred. However, Adam Sanborn and Nick Chater, cognitive scientists at the University of Warwick, have shown that the brain does not need to be a perfect Bayesian computer to construct perceptions from visual stimuli. It just needs to approximate one.[3] They've suggested that the brain is a "Bayesian Sampler," meaning that for each perceptual judgment, the brain only needs to consider the probabilities of a few alternatives. This is vastly more efficient than considering the entire universe of possibilities. In the crater illusion, the brain only considers two possibilities: concave or convex. There are certainly more interpretations than that, for instance, what it really is, namely, shades of gray on a flat surface. That our brains do not immediately jump to the true interpretation speaks to the power of prior experience.

There is no straightforward solution for how many possibilities the brain need consider at any given moment. The neural data suggests that the more ambiguous the input, the more possibilities the brain will consider in its interpretation. This is why two people can witness the same series of events, or read the same material, and come to wildly different conclusions. Each individual will consider different possible interpretations of sensory stimuli, and these possibilities will depend on their own past experiences and probability estimates.

Importantly, none of this need be conscious. The brain, like every other organ in the body, consumes energy. It will try to do its job using the least amount of energy necessary. This means that the possible interpretations of a sensory stimulus will tend to follow well-worn, energy-efficient pathways. The crater stimulus looks the way it does because we have experience with objects that resemble that shape. Think of candies, blister packs, push buttons, and clothing snaps. Only graphic artists might see shaded circles at first. None of these possibilities actually reach consciousness until the winner takes hold of the mind. According to Bayesian theory, only the most likely one reaches awareness, and only one interpretation can occupy that slot at a time.

Visual illusions are the easiest way to demonstrate the malleability of our perceptions, but they are not the only sensory modality subject to Bayesian inference. Tactile illusions are plentiful, too. And some argue that the sense of touch is the very foundation of personal identity, for touch demarks the boundary between the body and the external world.

Vincent Hayward, a professor at McGill University in Montreal, has studied tactile illusions for years and has cataloged the different types.[4] Some require specialized equipment to demonstrate, but others can be done with household items. Perhaps the easiest, although least consistent, is the Aristotle illusion. With your eyes closed, touch the end of your nose simultaneously with your index and middle fingers. Now, cross your fingers and do the same. For many people, the crossed-fingers condition feels like touching two different surfaces.

Another illusion is the distance misjudgment. For this, unfold a paper clip and position the ends about an inch apart. With eyes closed, touch the ends to the palm of your hand. You should feel two points. Now touch the ends somewhere on your lower leg. It will feel like one point. This may seem strange unless you know that the brain has a denser representation of the surface of the hand compared to the leg. Certain parts of the body—the hands, the face—feel the world in high resolution, while other parts encounter things in lo-res. Like visual illusions, tactile illusions occur largely outside of awareness. You feel something and your brain interprets the sensation as caused by what it thinks is the most likely event. It uses a lifetime of experience touching things and being touched to build up a mental representation of what the world feels like.

The final illusion is not really an illusion, but it is central to the construction of self. For this, you need a partner. Again with eyes closed, use your index finger to lightly touch your cheek. Now have your partner touch your cheek with their index finger (it helps if both of your fingers are warmed to the same temperature). Do the touches feel different? Of course. In one case you are touching yourself, and in the other, someone else is. When you touch your cheek, do you feel it in your cheek or your finger? The answer is both. But when someone else does it, you only feel it in your cheek. In the self-touch condition, the brain creates a prior representation of what it is going to feel, and then compares the actual sensation to what it expects. This is a form of Bayesian inference in which the act of touching creates the prior probability. You already have a good idea of what your cheek and your finger feel like. When someone else does it, though, your brain does not have good priors. You can simulate the effect of other-touch by reaching over the top of your head and touching the opposite cheek. Why does this feel different from the regular way? The brain's most probable interpretation of self-touch is the right hand touching the right cheek and the left hand touching the left cheek. When you cross, you create a new configuration for which your brain doesn't have a good prior. So it feels different.

Interestingly, when researchers in Sweden conducted a self-touch versus other-touch experiment using fMRI, they found that self-touch caused widespread deactivation in sensory regions of the brain that did not occur during other-touch.[5] Beyond the cortical regions, these changes were observed in the thalamus and brainstem, suggesting that the modulation—dulling, in the case of self-touch—of sensory experience may extend all the way to the spinal cord. This is a startling result. It means that your expectations of the world, which are shaped by your actions and the schema that have developed around them, change what you feel at the level of the nervous system where the information comes in.

These visual and tactile illusions are more than parlor tricks. They illuminate how sensory input does not allow an exact reconstruction of the world. There are always multiple possibilities for the cause of every sensation. In order to sift through the possibilities, the brain must use both prior experience and predictions from motor action to craft a best guess for what it is experiencing.

The present-self exists for just a second or two, but even that fleeting existence can't be separated from the compressed schemas in our heads. Present-self uses prior knowledge to interpret what is happening so that it can get ready to predict what comes next. This means that there are always multiple possible interpretations. And though the notion of the Bayesian brain is usually thought of in terms of perception of the external world, it also applies to the perception of our own bodies, which is called interoception. And if there are multiple interpretations of the sensations of our own bodies, then, as we shall see in the next chapter, there is the possibility for the construction of multiple identities.

CHAPTER 5

The Bayesian Self

H OW DO WE PERCEIVE OURSELVES? THE QUESTION IS central to the construction of personal narrative and ultimately who we think we are. Self-perception shares many of the same processes as external perception: backward-looking schemas to interpret what we are feeling and forward prediction to decide what we are going to do. And just as for external perception, there are multiple interpretations of internal feelings. This means there are multiple potential yous. Within this multiverse of yous lies the enticing possibility for controlling, even altering, the perception of who you are.

We have already established that although the brain is the organ that constructs the sense of self, it has no way to include itself in the construction. But if the brain has no perception of itself, then where do we locate the self? You might think, logically, that you are in your brain, and therefore your head. Others might more symbolically focus on their heart. Indeed, when people are asked directly to locate their self, three major clusters emerge: one between the eyes, one around the mouth, one in the center of the chest, and a minor one in the abdomen.[1] However, the location of self is more nuanced than that, seemingly varying based on one's mood.

In 2018, researchers in Finland dissected the location of self by asking 1,026 participants to identify the location of one hundred feelings on an outline of a body.[2] They ranged from negative emotions like anger, shame, guilt, and exclusion to positive emotions like happiness, laughing, success, and closeness. The researchers also included cognitive states like thinking and reading as well as unwell states like feeling nauseated or having a cold. A remarkable map emerged.

The map of emotions represents a probability map. For a particular emotion, say, gratefulness, it gives the probability of feeling it in a particular part of the body, say, the chest. In Bayes's terminology, this is the likelihood of a feeling in the chest given that you are grateful. But we can flip the map around and ask about the posterior probability: Given a feeling in your chest, what is the probability that it exists because you are grateful?

The problem of posterior probability is that many emotions are associated with a feeling in the chest. This is the conundrum of internal perception. Roughly half of the emotional states included in the survey were associated with a feeling in the chest. This is where past experience influences self-perception. Consider a person who rarely feels gratitude. If they felt a fluttering in the chest, it is unlikely they would attribute it to feeling grateful. Instead, they might interpret it as anxiety.

A version of this idea was first proposed in 1884 by the philosopher-psychologist William James and a year later by the physician Carl Lange. The James-Lange theory of emotion proposes that physiological changes in the body—such as heart fluttering—drive the perception of emotion. James is famously associated with the bear conundrum. You are walking in the woods

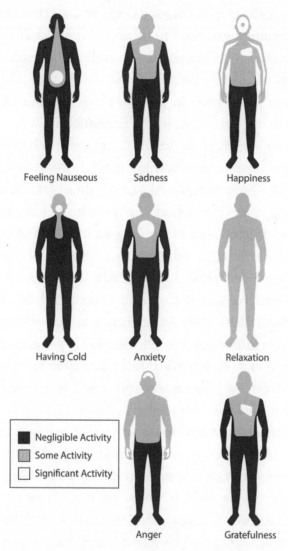

Locations of physical sensations associated with different emotions. *Adapted from Nummenmaa et al., "Maps of Subjective Feelings," 9198–9203.*

and happen to meet a bear. Without thinking, you start running. Of course, you are afraid. But are you running because you are scared? Or are you scared because you are running? James and Lange said it is the latter. Physiological changes cause you to be afraid, not the other way around.

Looking at the map of emotions, it is evident that both positive and negative emotions tend to be associated with feelings in the chest and abdomen. Although this correlation doesn't resolve the question of causality, it does raise the question of why emotions are perceived in the body in the first place. Is something actually happening there? Or is it all in the imagination?

The answer appears to be a bit of both. What these high-arousal states have in common is the release of hormones. Stress hormones like epinephrine—aka adrenaline—and cortisol have wide-ranging effects throughout the body. If you have ever had the unfortunate circumstance of needing an injection of epinephrine to treat a severe allergic reaction, then you know what a rush of pure adrenaline feels like. It causes the heart to beat faster and stronger, some likening it to a drum about to beat out of their chest. In the 1920s, when adrenaline was isolated and injected into human volunteers, some felt their heart pounded like the incessant percussion of Ravel's *Bolero*. Everyone trembled uncontrollably. And a few subjects were possessed of a disembodied fear to the extent they felt they might die.[3]

Because adrenaline drives real physical changes throughout the body, you feel its effects in your internal organs. But the brain still has to interpret these sensations like any other perception. Unlike external perceptions, though, we usually don't have a way to verify internal sensations. When someone says they have "butterflies in their stomach," we know it is a metaphor, but we have to take it on faith that they feel a fluttering of some sort. By contrast, a statement like, "that flower is a pretty shade of red" can be immediately verified.

The psychologist Lisa Feldman Barrett has gone as far as to argue that there are no universal emotions.[4] Instead, she says, a person constructs an emotion based on their interpretation of how they are feeling. These interpretations derive from both past experience and the available vocabulary a person has to describe different emotions. In other words, you can only experience emotions for which you have labels, which means you have to learn an emotion as opposed to it being innate. Another key feature of Barrett's theory is the lack of localization of emotions to particular parts of the brain.

Because the labeling of emotions with words is a central part of Barrett's theory, this would seem to deny emotions to any person or animal that can't speak. However, this doesn't jibe with commonsense observations of dogs and cats that clearly experience states of anxiety and fear, as do infants. The neuroscientist Jaak Panksepp put forward an alternative view of emotions.[5] Over the course of his fifty-year career, Panksepp studied the expression of emotions in animals and argued for the existence of seven basic systems in any animal: rage, fear, panic, lust, care, play, and seeking. Moreover, Panksepp thought each emotion resulted from activity within separate brain circuits and neurotransmitters. For example, care was associated with oxytocin, and rage with the neurotransmitters substance P and glutamate. He also thought that anyone who still believed in the James-Lange theory of emotion was an idiot.[6]

Given how central emotions are to the narratives we construct, it is helpful to consider why emotions evolved in the first place. Charles Darwin recognized that many animals seem to express emotions in ways similar to humans, suggesting that emotions serve a common function.[7] For example, think of hunger. Although it is not typically considered an emotion, hunger is a physiological state that causes an individual to reorient their behavior toward seeking food. Because hunger isn't associated with a specific motor action, it is not a reflex. Rather, hunger is like a course correction. Fear is similar but with the opposite effect, causing an individual to flee (or possibly freeze). Ralph Adolphs, a neuroscientist at Caltech, has promoted this view, advocating for a sort of middle ground between Barrett and Panksepp.[8] Adolphs's explanation is quite compelling because it assumes an evolution-ary value to emotions. As he says, "Emotions evolved to allow us to cope with environmental challenges in a way that is more flexible than reflexes but doesn't require the full flexibility of voli-tional, planned behavior."

I have spent a fair bit of space on emotions because they are central to our daily experiences, and for that reason alone, they play a critical role in who we think we are. If sequences of events constitute the stories we construct in our heads, emotions are the accompanying soundtrack. In many languages, the centrality of emotion is made literal by how we communicate our feelings. In English, we say, "I am happy" or "I am sad" or "I am in love." Of course, you could also say, "I feel happy," but that is not the same thing. When you have a feeling, you co-opt the tactile perception system and use Bayesian inference to classify what it is. But when you say, "I am," you are transformed. You become the emotion. And with at least a hundred different emotions, there could be a hundred different versions of a person.

Being versus feeling may seem like a grammatical nuance, but a simple exercise can demonstrate the difference. The next time you are angry or upset, instead of saying, "I am angry," say to yourself, "I feel angry." You should notice an immediate dif-fusing effect. To feel an emotion is a more neutral appraisal of your internal state. With some effort, you may even be able to isolate the feeling to its naked perceptual component. If you are prone to anxiety, this trick often works wonders. When you feel a fluttering in your stomach, instead of concluding, "I am anx-ious," focus on the sensation as if you were a doctor probing the characteristics of your body. Note what makes it better, and what makes it worse. Does it feel like hunger? Or something else? This technique requires you to detach or dissociate from your body. You might even imagine yourself floating above your body and looking down on it.

IN THE LAST CHAPTER, I MENTIONED THAT A PERFECT Bayesian brain would need to keep track of all the things that had happened to it. Neural data suggests that the more ambiguous the sensory input, the more the brain has to rely on past experience. But what happens when you have no experience and you misinterpret something? We would call that a perceptual error. These types of errors turn out to be central to how the brain builds up a representation of the world and—if we're referring to internal perception—a representation of your own body. They are called prediction errors.

Prediction is baked deeply into the brain. Up to this point, I have explained how prior probabilities affect the interpretation of sensory experiences. But simply interpreting incoming sensory data is not an effective way for an organism to move about the world. Such a creature could, at best, only react to events as they happened. Evolution stumbled upon a far better strategy: prediction. An organism that can make a prediction about the future— even if only for a second or two—has a tremendous advantage over less-forward-thinking animals. Successful predators, for example, predict where their prey will be, while prey that anticipate danger are more likely to escape.

Prediction also allows the brain to hone its estimates about the nature of sensory experiences. When you misinterpret something, your brain uses this information to update its beliefs about the world—namely, what caused the sensory input. Returning to the theme of first experiences, prediction errors are especially powerful because you have no prior expectations about how something should feel. Think again about your first romantic kiss. Although you would have had some expectations based on wide cultural exposure to kisses, you would not have had an actual sensory experience. Your expectations might have been based on the prince waking Sleeping Beauty; a kiss turning the frog back into the prince; the iconic Times Square kiss celebrating the end of World

War II, or many famous movie kisses (*Titanic, The Princess Bride, Gone with the Wind, Ghost, Brokeback Mountain*). But these are fictional. It is very likely that your actual first romantic kiss was nothing like any of these. More awkward than you'd expected? Sloppier? Too much tongue? These prediction errors result in an immediate revision of expectations for all future lip-locking sensory experiences.

Much of neuroscience research has focused on prediction errors in the context of anticipating things in the environment. This makes sense, as there is a strong evolutionary pressure to anticipate and avoid things in the world that might kill you. Similarly, there is an equally strong pressure to anticipate where to find good things like food and mates.

Yet the brain also uses prediction errors to interpret internal sensations about the body. It's another way that we construct a sense of self. To demonstrate, stretch out your arms to the sides. Then, with eyes closed, alternately touch your nose. Apart from being a crude field sobriety test, it demonstrates your sense of position. Specialized neural receptors, called proprioceptors, are located in your muscles and tendons. They send signals to your brain about the position of your body parts. To touch your nose, your brain uses that information to control your muscles so that, hopefully, your finger lands on your nose. You may notice that as your finger gets closer to the target, you tend to slow down your movement, and you may even experience a little trembling. This is your brain making course corrections. In order to do that, though, it has to know where the target is and form a prediction of where your finger is headed. Any deviation from the predicted course is an error that gets corrected.

That is a rather long-winded explanation for something you don't typically think about, but it's central to the sense of self. Moreover, the type of self I am describing is a core feature of all animals. The first, and most basic, task for any creature is to differentiate what it does to the world and what the world does to it. Some have called this a "minimal self," which is the feeling that a movement is performed by *me* or that an experience is *mine*.[9] When you pick up a coffee mug, you know that you are responsible for the action. The mug did not move to your hand. You know this without even looking at it. But how? The same way you touched your nose. Your brain created a trajectory for your hand to get to the mug. This is called a forward model, and it is uniquely responsible for your sense of agency. If the mug magically flew into your hand, apart from being very surprised, you would immediately recognize that it was not caused by any action of yours because your brain would not have a forward model.

As basic as the minimal self is, it can still be fooled. You may be wondering how you could possibly mistake an action you make for something else. In 1998, Matthew Botvinick and Jonathan Cohen, psychologists at Carnegie Mellon University, demonstrated what's now known as the "rubber hand illusion."[10] First, they had a person sit with their arm resting on a table. The subject's arm was hidden from their view by a screen. The researchers then stuck a rubber arm out from behind the screen so that the subject could see it. With the subject watching the rubber arm, the researchers stroked the real arm and the fake arm simultaneously with a soft-bristle brush. Eighty percent of the participants reported feeling the brushing occurring in the fake arm. The self had been extended to an inanimate object. The rubber hand illusion shows that our notion of self—even the minimal self—depends on the integration of multiple sensory inputs, especially vision and touch. These two senses are central to defining the boundary between the body and the environment. But when visual and tactile signals conflict, the brain does the best it can to make sense of such information, including the extension of self to the rubber hand.

The rubber hand illusion isn't just a neat science trick. The neural mechanisms behind it also apply to everyday things like automobiles. When you drive a car, especially one you have driven a long time, the vehicle feels like an extension of yourself. Normally, your personal space extends in a halo around your body, but when you are driving a car, it envelops the car itself.[11] The effect is not a result of simply being in a car, because you don't feel it as a passenger. Instead, the illusion arises because your brain makes forward models of how the car will react to your movements of the steering wheel, accelerator, and brake pedal. These forward predictions eventually become subconscious and that's what makes the car feel like part of your body.

The rubber hand illusion and the expansion of self in an automobile show that the sense of self is not a static concept. The self is dynamic, expanding and contracting as specific circumstances dictate.

So far, I have been referring to a sort of minimal self that processes streams of sensory information as they flow into the brain. That is you in the moment. The bigger question, which I raised earlier, is, How does this instantaneous present-you connect to yesterday's self or the self of your childhood? As we've seen in earlier chapters, the answer appears to be the construction of narratives.

A narrative connects a series of events. Your personal narrative is the story of your life, of everything that has happened to you. Subjectively, this notion of self feels different from the instantaneous one navigating rush-hour traffic. It is a self that extends in time, connecting present-day "you" to childhood "you," and it is the self that you project into the future. It is a *narrative self*. While all animals have a minimal sense of self, only humans have a narrative self because only humans possess the language to tell stories.

The narrative self sounds rather abstract, so it is no surprise that philosophers have been debating its form since at least the

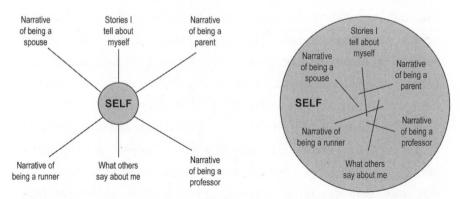

Two theories of the narrative self. Left: Dennett's model of a center of gravity. Right: Gallagher's interpretation of Ricoeur's model of a decentralized self. *Adapted from Gallagher, "Philosophical Conceptions of the Self," 14–21.*

eighteenth century. Scottish Enlightenment philosopher David Hume posited that the narrative self was a fiction, a sort of fairy tale we told ourselves to connect the moments in our lives. In the 1990s, philosopher Daniel Dennett argued that the narrative self was an abstract "center of gravity," a nexus for all the facets of you.[12] An alternative view, suggested by the French philosopher Paul Ricoeur, conceives of the narrative self as something distributed and decentralized.

Personally, I think the evidence tips in favor of Ricoeur's distributed narrative self. If the minimal self can expand and contract, then surely the narrative self can, too. It may seem impossible to disentangle these debates. In part, our intuitions fail us when thinking about ourselves, as I've shown in previous chapters. Neuroimaging, though, has begun to reveal what the brain looks like when it tries to think about the self.

There are many ways to think about the self. I've chosen to highlight emotions because they figure prominently in personal narratives, but they are also personal. Indeed, how one person experiences happiness may be different from how someone else does, even as both would recognize their own version as happiness when they experienced it. Similarly, tactile sensations—both internal and external—belong to the individual, and different people might experience them in idiosyncratic ways. But this is just the tip of the iceberg. You can think about yourself in an infinitude of ways. Needless to say, neuroscientists have been scanning people thinking about themselves in all of them.

What's illuminating are the commonalities. If you collect up and compare all the fMRI studies in which subjects had to think about some aspect of the self, a consistent picture emerges. The first time such an analysis was attempted, in 2006, the researchers identified a strip of cortex running down the midline from the front to the back of the brain that appeared to be independent of the specific modality invoked to think about the self.[13] What modalities? Everything from sensory to emotional to memory. These cortical midline structures—CMSs—are thought to be positioned in a way that they can link sensory systems that process physical sensations (e.g., the minimal self) to more abstract representations that rely on memory and symbolic representation (the narrative self). When the researchers looked to see whether there were any subdivisions within the CMSs that might be associated with certain domains of self, they couldn't find any. This finding, I would argue, is consistent with the distributed model of the narrative self proposed by Ricoeur.

The cortical midline structures also show up frequently in resting-state fMRI, or rs-fMRI.[14] As you might infer from the name, in resting-state studies, the participant doesn't do anything. They simply lie in a relaxed state while being scanned, typically for about ten minutes. The researchers then examine the patterns of brain activity to find out which regions are correlated with each other. The surprising result is that there exist patterns at all. When you're letting your mind wander, you might think that brain activity would flit all over the place. Quite the opposite. The cortical midline structures do, in fact, fluctuate in activity, but they do so in synchrony. When it was discovered, this pattern was dubbed the default mode network (DMN) because it seemed to be the pattern of activity the brain defaults to when it isn't doing something else. And not all of this activity is due to self-referential navel-gazing, because the DMN persists even under sedation.[15]

The takeaway is that the midline structures are always on and working in concert with each other to bind together all the systems that constitute the self. The degree to which they do that can be amplified by removing external distractions, leaving the mind to wander about the self. Conversely, the midline activity can be diminished by engaging in something that requires externally directed focus, or what the psychologist Mihaly Csikszentmihalyi famously called *flow*.[16]

AT THE BEGINNING OF THIS CHAPTER, I SUGGESTED THAT internal sensations, which are central to how a person feels about themselves, are Bayesian predictions no different from external perceptions. Because internal feelings can map back to multiple emotions, the brain uses prior probabilities to determine which emotion you're actually experiencing. This may sound backward, but this would explain the difficulty we often have in figuring out how we feel in the moment. Moreover, the brain regions responsible for this function are centrally positioned to integrate a wide variety of sensory and memory inputs.

So, what? You may be wondering whether these findings have any practical implications for our lives. Indeed, they do. To some degree we can learn to choose how we interpret the internal feelings we've come to think of as automatic.

While your Bayesian brain constantly makes its best guess for what you're experiencing—internally and externally—it also collects feedback about how well it is doing. When your brain misinterprets a sensation, it generates a prediction error, which is a signal that something must be corrected. The prediction error can be used to improve the brain's guess the next time around. But this is not the only function of a prediction error. Your brain could just as easily change the sensory inputs to conform to its own predictions.[17] This is, quite literally, a case of seeing what you want to see, or bending the world to conform to how you imagine things are.

In 2015, a viral internet meme known as *The Dress* illustrated how far this could go. It began with a Tumblr posting of an image of a person wearing a lace-trimmed dress made of horizontal bands of colored material. Some people saw the bands as white and gold. Others saw them as blue and black. Scientists became fascinated with the phenomenon and launched several studies to understand this apparent disparity in opinion. One study found that 57 percent of people saw the dress as black and blue, 30 percent as white and gold, and 11 percent as brown and blue.[18] More interesting, though, was that people's perceptions could be nudged around by changing the background color that the dress appeared against. Just as Bayesian probability predicts, a person's perception of the dress is affected by their prior assumptions about the lighting conditions—sunlight versus fluorescent. The researchers could modify the assumptions by drawing attention to a visual cue that implied different lighting conditions. This works because visual perception—like any perception—is an active process. You decide what to look at. The researchers showed that by looking at different parts of the image, the perception could be changed. Flip it around, and you can choose which parts of the image to focus on so that it reinforces your preexisting prediction of the color of the dress.

If such a maneuver works for something as basic as visual perception, then it can certainly work for internal feelings. Perhaps even more so because you don't have any external, shared truth to compare your impressions to. For example, if you feel anxious, you have the choice to interpret the sensation as anxiety or, by focusing on different aspects of the experience or by changing your breathing, you can change the sensation to conform to a different interpretation. This idea of agency is foundational to cognitive behavioral therapies. But why stop there?

If sensations can be altered to conform to different interpretations, so can the entire perception of self. To do so requires a conscious shift in the ongoing narrative of you. A plot twist, if you will. Taken to the extreme, it might even be possible to become a different person. In the next chapter, we'll learn about how—and why—we deploy different versions of ourselves under different circumstances.

CHAPTER 6

All the Yous

A T THE BEGINNING OF THIS BOOK, I INTRODUCED THE idea that everyone has three versions of themselves, split along the temporal dimension: the past-, present-, and future-yous. In fact, there exist more. We all carry around different versions of ourselves in our heads. There are the versions that we adopt in specific social situations that, in turn, differ from the one that we hold in private, when we are alone. By this point you may recognize that the question of which one is the "real" you is a bit of a red herring. Thanks to humans' ability to dissociate, all of them are real.

The notion of split selves is not new. Sigmund Freud famously split the psyche into three parts: the id, the ego, and the superego. Later, Carl Jung argued that everyone carried a "shadow," that is, a dark side to their personality that could temporarily overwhelm the conscious side. But neither Freud nor Jung thought that these parts of one's personality were fully formed in their own right. Like an iceberg, the bulk of an individual's personality was believed to lie beneath the surface of consciousness. By the late nineteenth and early twentieth centuries, however, a growing number of psychiatrists came to believe that some patients could harbor multiple distinct, yet complete, personalities.

The idea of multiple personalities continued to grow throughout the twentieth century, in large part as a result of a handful of books and movies that captured the fascination of the public. Meanwhile, most psychiatrists did not know what to make of multiple personality disorder (MPD). They asked: Was it real? How should it be treated?

We still don't have clear answers to these questions, but the evolution of our understanding of MPD, now called dissociative identity disorder (DID), holds important clues for the fluid nature of who we think we are.

THE NOTION OF AN INDIVIDUAL POSSESSING MULTIPLE PERsonalities broke into popular consciousness in 1906, when Morton Prince, a neurologist in Boston, published a book-length account of one of his patients.[1] The book opens with this description:

> Miss Christine L. Beauchamp [a pseudonym pronounced Beecham] is a person in whom several personalities have become developed. In addition to the real, original self, the self that was and which she was intended by nature to be, she may be any one of three different persons. I say three different, because, although making use of the same body, each, nevertheless, has a distinctly different character; a difference manifested by different trains of thought, by different views, beliefs, ideals, temperament, tastes, habits, and memories. Two of these personalities have no knowledge of each other or the third. The personalities come and go in kaleidoscopic succession, many changes often being made in the course of twenty-four hours.

For better or worse, Prince created a terminology to describe the fracturing of consciousness that has become part of our modern psychological landscape. Presciently, Prince did not like the term *multiple personality*, preferring "disintegrated" to indicate that secondary personalities were only part of the original, whole self. He also dubbed them alternates, or *alters*, as people later began to call them.

Miss Beauchamp did not originally present herself to Dr. Prince complaining of multiple personalities. Rather, when she was twenty-three, she came into his care to be treated for "neurasthenia," a vague term encompassing a constellation of physical symptoms. In Miss Beauchamp's case, they included headaches, insomnia, body aches, fatigue, and poor nutrition.

Psychiatrists of the era had few tools to treat their patients. Pharmacologically, there were sedatives, primarily morphine and chloral hydrate. Both drugs had become mainstays of late nineteenth-century asylums because they were extremely effective in calming agitated patients and putting them to sleep. By the early twentieth century, they had crossed over into daily use by people suffering from routine anxiety and insomnia. The effects of morphine are well known, common as it still is today. Chloral hydrate, meanwhile, has not been in common use since at least the late 1990s. A century ago, it was the original roofie. When chloral hydrate was mixed into an alcoholic drink, it was called a Mickey Finn—as in "slipping someone a Mickey." A highly effective concoction for knocking someone out and leaving them without memory for whatever events followed.

Prince did not document whether he tried morphine and chloral hydrate, but we assume he did, because he noted that Miss B. had failed to respond to "conventional methods." Then, on April 5, 1898, he hypnotized her. According to his notes, the procedure was repeated over several days, and she immediately began sleeping better and eating with a ravenous appetite. Over the ensuing weeks, however, Prince began to discern what he thought were different personalities under hypnosis. The first to appear called herself Chris, shortened from Christine. Chris later changed her name to Sally and had the peculiar affectation of referring to herself in the third person.

Over time, Prince became convinced that Sally was a distinct personality from Christine. Christine was anxious and morose, but Sally was bold, vivacious, and possessed of a "saucy deviltry." Sally stuttered, while Christine did not. Sally did not like Christine, calling her a "stupid chump" and smoking a cigarette to leave a bad taste in her mouth for Christine to experience.

If this all sounds faintly reminiscent of *The Strange Case of Dr. Jekyll and Mr. Hyde,* that may not be a coincidence. Robert Louis Stevenson had published his novel in 1886, thirteen years before Miss Beauchamp presented to Dr. Prince. No doubt she would have been familiar with the story. Today, though, Stevenson's central message seems to be less widely appreciated. Mr. Hyde was not simply an alternate, evil personality. Jekyll recognized, and struggled with, the duality of his own nature, as two captains might fight for control of the helm of a ship. To the public, he came across as a serious, moral doctor, but privately, he wanted to drink deeply of life's pleasures. His potion let him do that without the inconvenient shackles of shame and guilt. Jekyll knew when he was Hyde. Some would argue that he wanted to be Hyde all along.

Like Dr. Jekyll turning into Hyde even without his potion, Miss Beauchamp eventually began transforming into Sally without hypnosis. Sally would play practical jokes on her other self, like writing letters committing herself to engagements that Miss Beauchamp would never go to. After a year of this, a new personality appeared, which Sally called "the Idiot" because the new personality knew nothing of the other two. After several years of tug-of-war, Prince reported that through repeated hypnosis sessions he was eventually able to merge the personalities of the original Miss Beauchamp and the Idiot. But this required exorcising Sally, banishing her "back to where she came from."

There's a long history of such stories of transformation, especially under the power of potions. In the vein of Jekyll and Hyde, Bruce Banner becomes the Incredible Hulk after being exposed to high levels of gamma radiation. Subsequently, every time Banner gets angry he transforms into the Hulk. Although he doesn't transform to quite the same degree, Peter Parker becomes Spider-Man after being bitten by a radioactive spider.

The fact that many of our most popular superheroes originated in some kind of magical transformation speaks to the power of this narrative trope. The reason is simple: stories of transformation tap into our innate desire to assume different identities ourselves. Such transformations allow us to compartmentalize different versions of ourselves that maintain different identities.

THE STORY OF MISS BEAUCHAMP OFFERS TWO INSIGHTS into how we construct and deconstruct personality. The first is that all of us have different personas that coexist within the umbrella of "our" personalities. Though it may seem implausible, I suspect Miss Beauchamp (or Sally or the Idiot) truly believed they were different people. Her perception of self was altered under different circumstances. But isn't it the case that we all have slightly different perceptions of ourselves based on the social milieu? There is the version of you at work, a version with friends, the one with family, itself different with parents, siblings, and children. And then there is the solitary you.

The difference between Miss Beauchamp and the rest of us appears to be a matter of degree. Like Dr. Jekyll and Mr. Hyde, Christine Beauchamp had different names for her alters. Granting the possibility that we have a range of personalities within each of us, it may seem extreme to give each a different name, but this is, in fact, common. They're called nicknames. For example, my professional names are: Doctor Berns or Professor Berns, depending on whether I'm in a clinical or an academic setting. My writing name is Gregory, but my friends call me Greg. As a child, it was sometimes Bernsie, or the one I hated most: Curly, for my hair.

To some extent, we maintain these identities as if they are on separate trajectories. This separation was made hilariously clear in a classic *Seinfeld* episode. George has been going out with a woman named Susan for some time, when Elaine, at Jerry's suggestion, invites Susan to join her at a show. George is furious. In his mind, he has separated the world with Jerry, Kramer, and Elaine from the world he inhabits with Susan. "Independent George," his bachelor persona, lives in the first one, his paired self in the second. By breaching that separation, Jerry has threatened George's universe of multiple worlds. And as everyone knows, when worlds collide, they explode, and Independent George will die.

The second insight is that Miss Beauchamp and Dr. Prince together created the template for multiple personalities. Consider the manner in which Christine Beauchamp's alters appeared. There's no evidence that there was a trauma that fractured her personality. And as far as we know, they did not exist until Dr. Prince hypnotized her. A strong case could be made that Prince created her other two "separate" personalities through the power of suggestion, launching what would eventually become the most controversial disorder in the annals of psychiatry.

The idea of multiple personality disorder, however, did not arise from nowhere. For much of the twentieth century, patients, invariably women, suffering from a menu of vague physical and mental symptoms, would land in psychiatrists' offices. Multiple personality was not considered a distinct disorder but rather a symptom of hysteria, which also included fugue states and amnesia, convulsive attacks, motor paralysis, anorexia, and anesthesia—meaning insensitivity to pain.[2] The term *hysteria* derives from the Greek word for womb, reflecting the ancient belief that these ailments were caused by a wandering uterus. Although by the twentieth century psychiatrists no longer believed the uterus went on a walkabout, hysteria was still viewed as a women's problem.

Freud is commonly credited with developing the leading treatment for hysteria, whereas the symptoms and treatment of multiple personality disorder / dissociative identity disorder trace their origin to one of Freud's peers: Jean-Martin Charcot.

One of the most influential neurologists in the history of medicine, Charcot practiced at the famous Salpêtrière Hospital in Paris from the 1860s through the 1880s. During that time, he was among the first to describe multiple sclerosis and what later was called Parkinson's disease. He pioneered the standardization of the neurological exam, which consists of the systematic and precise examination of sensation and reflexes. Coupled with his knowledge of anatomy, Charcot's exam allowed him to localize neurological problems without the use of modern radiographic imaging. The neuro exam is still the starting point for all neurological consults.

Freud was undoubtably influenced by Charcot's work, but the two disagreed fundamentally about the root cause of hysteria. Charcot thought it was a neurological problem caused by a traumatic event, and a cure could only be achieved through hypnosis. Freud believed that hysteria was a psychosexual problem that could be treated through psychoanalysis.

Now, a century later, neither view seems particularly relevant. In part, that's because patients are no longer diagnosed with hysteria. But that does not mean the syndrome the term described has gone away. The body manifests physical symptoms for all sorts of reasons, and when so-called organic causes are ruled out or cannot be found, the patient with unexplained physical ailments is said to suffer a functional illness, like irritable bowel syndrome or chronic

fatigue syndrome. These are real syndromes with real physical suffering, but their pathologies are deeply tied to the nervous system.

What's relevant for the conception of self is how the early
descriptions of hysteria led to the popularization of MPD. Like
everything in psychiatry, Freud was a major influence. In the period between the stories of Dr. Jekyll and Miss Beauchamp, Freud
teamed up with Josef Breuer, another neurologist, and published
Studies in Hysteria. The entire volume of case histories would become foundational to the psychoanalytic movement, but the story
of Anna O. stands out especially. She became the archetype for
the so-called cathartic cure.[3]

Anna O. was Breuer's patient, and her story was told through
his perspective. In Breuer's telling, Anna O. was a smart, attractive twenty-one-year-old Viennese woman born to a relatively
wealthy family. Her problems began when her father, whom she
adored, fell ill with tuberculosis. For months, Anna tried to nurse
him back to health, but she herself became exhausted and was ultimately forced to stay away. She also began to have coughing fits
in between long bouts of sleeping. As the months went on, Anna
began to suffer from a perplexing array of physical symptoms. She
had problems with her vision. She developed intermittent right-
sided paralysis. Parts of her body went numb, and she had trouble
speaking. Breuer writes that her personality split into a relatively
normal, albeit sad, person and a morbidly foul, agitated person
obsessed with hallucinations of black snakes.

When her father died, Anna completely broke down. After two
days in a deep stupor, she awoke and no longer recognized her family. She knew only Breuer. She refused to eat unless Breuer was
present, and her condition deteriorated when he had to leave town
for a week. When he returned, Breuer could only revive Anna
through hypnosis.

These hypnotic sessions became her "talking cure," or what she
called "chimney sweeping." Between sessions, she was wracked with

such anxiety that she could be calmed only by heavy doses of chloral hydrate. According to Breuer, Anna's personality eventually fractured completely. She still alternated between the relatively normal and sick personas, but now the sick Anna believed she was living exactly one year earlier than her healthy counterpart. Breuer continued the hypnosis sessions for six months, targeting each of Anna's symptoms to bring about a catharsis. By June 1882, two years after her symptoms began, Breuer claimed that Anna had recovered. Three years later, when he and Freud published their account, Anna was said to be in good health.

And that is where Anna's story stood for seventy years. She had become the prototype of cathartic talking cures. Yet unfortunately it wasn't true. Even Freud knew this, reportedly telling Jung in 1925 that Anna had not been cured.[4]

In 1953 it was revealed that Anna's real name was Bertha Pappenheim.[5] It wasn't difficult to unearth details of Bertha's life, because she was already well known in the Viennese Jewish social scene. In the 1880s, when she was presumably suffering a nervous breakdown, Bertha was deeply involved in social causes. She traveled throughout the Baltics, ferreting out centers that trafficked Jewish women forced into sex slavery. It is hard to reconcile this version of Bertha with the woman Breuer described. A bit of medical detective work by the Canadian psychiatrist Henri Ellenberger uncovered the original records of Bertha's treatment at a sanitarium outside of Vienna. In contrast to the talking cure claimed in the published account of Anna O., it was discovered that Bertha had been heavily sedated with high doses of chloral hydrate and morphine for facial tics.

It's become clear that Anna O.'s story is ultimately a familiar one: a woman whose life and experience were co-opted to suit the agenda of some powerful doctors. Freud framed her as a poster child for the talking cure, but we will never know the true cause of Anna's symptoms. Although she was cast as the prototype of hysteria, if she were seen today, a possible diagnosis might be temporal lobe epilepsy. Anna's story was distorted to fit the pre-determined narrative that Freud and Breuer wanted to promote. Unfortunately, that narrative persisted almost a century, shaping both the psychiatric treatment of hysteria and the public's perception of multiple personality disorder.

ALTHOUGH THE CASES OF ANNA O. AND MISS BEAUCHAMP laid the foundation for the popular understanding of hysteria, when they were first published they were relatively unknown. The stories of Eve White in the 1950s and Sybil in the 1970s, however, landed like bombs in the popular press. Until then, hysteria and its attendant fracturing of personalities were the domains of cigar-smoking psychiatrists. After Eve and Sybil entered the lexicon, they made MPD seem as ubiquitous as a nervous breakdown.

The story of Eve White began like the others before her—with a referral to a local psychiatrist for treatment of physical symptoms, namely, severe headaches. Eve lived in rural Georgia, not far from where the famous Masters Golf Tournament was held. Psychiatrists were few and far between, so Eve had to travel to the Medical College of Georgia in Augusta, where she saw a robust, tow-headed psychiatrist named Corbett Thigpen. It wasn't the headaches that made Eve's case interesting to Thigpen. It was the blackouts that followed them. To help Eve recover the memories of what happened during these somnambulistic fugue states, Thigpen did as Breuer had done with Anna and Prince with Miss Beauchamp. He hypnotized her.

Thigpen described what happened next.[6] "The demure and constrained posture of Eve White melted into buoyant repose. With a soft and surprisingly intimate syllable of laughter, she crossed her legs." Thigpen found them attractive. He applied the standard psychoanalytic explanation of countertransference, in which the therapist uses his own feelings as a gauge for which emotions the patient is trying to elicit from him. Put simply, Thigpen attributed his new attraction to Eve herself. As he wrote, "Instead of that retiring and gently conventional figure, there was in the newcomer, a childishly daredevil air, an erotically mischievous glance." She called herself Eve Black.

Eve's story bore uncanny similarities to that of Miss Beauchamp. Whereas Eve White was married, Eve Black claimed she wasn't. And she liked to party, apparently disappearing for days, "seeking the company of strangers." Of course, Eve White claimed no memory of Black's actions. Thigpen and his colleague Hervey Cleckley even evaluated Eve's personalities using various standardized tests, including measuring her brain wave patterns with electroencephalography. Although the electroencephalograms (EEGs) were inconclusive, White scored a slightly higher IQ than Black. Over the fourteen months that Thigpen and Cleckley treated Eve, they logged one hundred hours of therapy, largely with hypnosis. By the end of this period, a third personality had emerged, calling herself Jane.

Eve's story might have stayed within the annals of psychiatry, except that Thigpen and Cleckley presented her case at the 1953 convention of the American Psychiatric Association, which was also attended by members of the press. When Eve's story hit the wires, the public was rapt. Thigpen and Cleckley convinced Eve to sign over the rights to her story to them, and their 1957 book, *The Three Faces of Eve*, became a best seller. The same year, Joanne Woodward starred in the movie version and won the Oscar for Best Actress. In these versions, the Eves disappeared and Jane lived happily ever after.

As fantastic as Eve's story was, the tale of Sybil took the salaciousness up a notch.

Sybil, whose real name was Shirley Mason, had suffered from a panoply of physical ailments and blackouts since her childhood in Minnesota in the 1950s. An only child of Seventh-Day Adventists, Shirley spent much of her youth playing with dolls and imaginary friends, even though her mother viewed such fantasies as the devil's work.[7] At the suggestion of a family practitioner, Shirley was referred to Connie Wilbur, a psychiatrist in Omaha, Nebraska. Wilbur had ambitions of her own. Women doctors were uncommon, and as a midwestern psychiatrist, Wilbur was stuck in a professional backwater. Everyone knew that the hotbed of psychoanalysis was New York. She took a liking to Shirley, who was uncommonly smart compared to her usual patients. Shirley, in turn, idealized Dr. Wilbur as the type of sophisticated woman that she hoped to become someday. But Wilbur had already made plans to leave Omaha to advance her career on the East Coast. Their therapy sessions ended after six months. It was only by chance that Shirley's and Wilbur's paths would cross again nine years later.

It isn't clear how Shirley located Dr. Wilbur in New York, but it wouldn't have been difficult, because Wilbur had established a highly successful practice combining psychotherapy with the use of new, powerful sedatives. These drugs, mostly barbiturates, such as thiopental and amobarbital, had gained popularity in both psychiatric circles and criminal investigations as "truth serums." She prescribed Shirley a cocktail of sedatives, including Seconal for sleep and an aspirin-amphetamine mixture for menstrual pain.[8] After a few months of living on uppers and downers, Shirley showed up at Dr. Wilbur's office claiming to be a little girl named Peggy. The next week, she arrived as Vicky, who said that there were, in fact, two Peggys—Peggy Lou and Peggy Ann.

Wilbur wanted to know what was happening when Shirley went on her rambles through New York as these different personalities. The only way she could learn was through a Pentothal interview. They began regular sessions, which always started with a syringe full of the barbiturate injected directly into Shirley's vein. Wilbur tape-recorded everything.

All of this would have remained privileged information between patient and therapist, but Wilbur had other plans. After accumulating boxes of notes and reels of tape recordings through the sixties, Wilbur approached Flora Schreiber, a writer, about turning Shirley's story into a book. She introduced Flora to Shirley, who was eager to get her story out to others who suffered from similar problems.[9] In the final version of the book, they changed Shirley's name to Sybil, which became the title that hit bookstands in 1973. It became an instant best seller. A movie for television aired in 1976 with Sally Field starring as Sybil, and, in a role turnabout, Joanne Woodward played the psychiatrist.

THESE STORIES OF MPD, WHILE FASCINATING IN THEIR OWN right, are extreme examples of a capacity that exists in everyone. Psychiatrists now use the term *dissociation* to describe mental states in which a person feels like they are not themselves. Dissociation may take the form of out-of-body experiences such as floating above oneself or seeing events from a third-person perspective or, at its extreme, a complete fracturing of the psyche into different personas. The ability to dissociate holds important clues to the construction of self because it demonstrates the fluidity of our memories. Everyone has memories from a third-person perspective, which, if you think about it, is impossible because you can't see yourself.

One of my most vivid third-person memories is of a traumatic experience from my youth. When I was sixteen, I rode my bike everywhere. Even though I was allowed to borrow my parents' car, I preferred the freedom of cycling. One day, I was riding along an access road that paralleled the interstate when a semi, in the rightmost lane of the highway, veered onto the shoulder. My mind could not understand what was happening as the truck plowed through the chain-link fence and headed straight for me at sixty miles per hour.

That's the point at which my memory fragments into a third-person perspective. As I remember it now, I am hovering above and behind myself, for I can see myself on the bike. I can picture the terror in the driver's eyes as he swerves at the last minute to avoid hitting me. The truck jackknifes in slow motion as the cab slams into a hill right in front of me, raising a massive cloud of dust. After a few minutes, it begins to clear and I can make out two people ejected onto the hill. I see myself running over to them. They're both alive, moaning in pain. But I can do nothing except comfort them while other people arrive. The driver says he is thirsty, so I give him my water bottle. I didn't know it at the time, but he was in shock, probably bleeding internally.

A man comes up to me and says, "Boy, I saw everything from up on the hill. I thought you were a goner."

Eventually, the paramedics arrive and take the two people away. I never found out what happened to them.

What I have described is an episode of depersonalization. The *Diagnostic and Statistical Manual of Mental Disorders*, Fifth Edition (*DSM-5*), describes it as "an experience of unreality, detachment, or being an outside observer with respect to one's thoughts, feelings, sensations, body, or actions." Depersonalization is common. In a survey of rural North Carolina in the summer of 1995, 19 percent of the population had experienced an episode of depersonalization in the past year.[10] Derealization, which is similar to depersonalization and characterized by experiences that are "dreamlike, foggy, lifeless, or visually distorted," occurred in 14 percent of the people in the same survey. Time dilation is a type of derealization. Although depersonalization and derealization have some aspects in common with MPD (or DID, in modern nomenclature), they are not the same. When you are depersonalizing, you still know that you are you, albeit from a third-person perspective. MPD is a complete split from who you are.

Although a traumatic experience is commonly thought to be the inciting event for the splitting of personalities, some researchers have questioned this etiology. The alternative posits that MPD and DID are the result of social learning and cultural expectations.[11] In the sociocultural theory, some people split their personalities having heard about the stories of Eve and Sybil. Indeed, the number of reports of DID and MPD increased dramatically after the publication of *Sybil*. To make matters worse, well-meaning therapists learned to cue their patients, probing for repressed memories and "parts of you with whom I have not spoken."

When considering these diagnoses, it is natural to question whether MPD/DID is real. After all, such patients seem far outside the realm of our normalized experience. But this is the wrong question. One out of five people will have an episode of depersonalization every year, which means that almost everyone will experience some type of dissociation in their lives. MPD/DID is the most extreme form it can take, but most people do not suffer such deep and lasting fragmentations of personality, even after a traumatic event.

We are left, then, with a spectrum of dissociative experiences ranging from the occasional, and common, depersonalizations to the rare full-fledged personality splits. Where you fall along the spectrum depends, in large part, on what you have been told about these types of experiences. If you had learned that everyone has out-of-body experiences, then your interpretation of depersonalization would be different than if you had never heard of it or, worse, you had learned of it through a sensationalized account like *Sybil*.

If we accept that dissociation is a normal part of the human experience, then we must also accept that it is a fabrication of the mind. There is no way to exit our bodies and observe ourselves. So why does it seem so real?

First, the memory of the experience is not the same as the actual experience. The actual experience occurs only once and then it passes into memory. As we've learned, every time it is remembered, it must be reconstructed, like a movie editor stitching together frames. Every time it is reconstructed, it gets corrupted a little bit by whatever is happening at that instant. In my case, the first time I recalled the experience of biking on the access road was when the guy came up to me and said he thought I was a goner. It was only minutes after the actual event, but the key phrase was when he said he "saw everything from up on the hill." At those words, I imagined witnessing what had happened from his point of view. I was still jacked up with adrenaline, so that first remembrance got mixed in with the actual events. I replayed the events even as they were still happening, so it's no wonder the imaginary third-person memories got mixed in with the first-person ones. Forty years later, now I can't tell the difference between my own experience and what was the result of that first imagined reexperience.

The ease with which we dissociate may also have something to do with the narratives that permeate our lives. It is difficult to think of oneself in multiple narratives at the same time. Narratives are constrained by space, and you can be in only one place at a time. Dissociation frees a person from this constraint. If I am floating above my body, observing myself from a different vantage point, then I have mentally broken the constraint of space by occupying two places simultaneously. If a person can dissociate into different points of view, then that is the same as saying that there are different narrators in their head. The degree to which a person can dissociate, or parcel out different perspectives on their experiences, will determine their propensity for keeping track of multiple narratives.

Earlier I mentioned the popularity of transformation stories, es-pecially for superheroes. Consider Batman, who lives a straight-up life as Bruce Wayne. He doesn't even need a magic potion to turn into Batman, his Mr. Hyde. Nor does Superman, who masquerades as Clark Kent. In a sort of rags-to-riches version of superhero trans-formation, Tony Stark—aka Ironman—is particularly appealing because he invented his transformation himself.

Once you start looking for transformation stories, you see them everywhere. They have become a culturally acceptable way to adopt different identities. Look at the popularity of cosplay and role-playing games. Now, one might argue that people know these identities are just fantasy and these activities are just en-tertainment. But are they? Many people who attend gatherings like Comic-Con or Burning Man spend much of the rest of the year planning for the event. This raises the interesting question of whether it's their day job or the festival that defines their sense of self. Or both.

Up to this point, I have been building a case for how our brains use narratives as templates for interpreting the world we live in. These narratives come from our own experiences, but they are overlaid on the stories that we have heard beginning in childhood and then throughout our lives. In the next chapters, we will con-sider some very common narrative forms that appear repeatedly in the stories we consume. We cannot help but graft our own ex-periences onto these ubiquitous narratives. Tales of multiple per-sonalities are variations on some of these common themes, and as I have suggested, everyone is on a spectrum of dissociation. To varying degrees we all take on different identities. Like channel surfing the different narrative streams we hold in our heads, each track links a different trajectory through past, present, and future.

CHAPTER 7

The Evolution of Narrative

B Y NOW, IT SHOULD BE APPARENT THAT OUR BRAINS trick us in many ways. The brain's bag of tricks includes the ability to dissociate and adopt a third-person point of view. It includes compression, the capacity to store memories in a space-saving format that lets us recall episodic memories. And finally, it includes the sneaky ability to shift time. Even when we think we're living in the moment, we're really processing information that has already passed. This final ruse has led to the evolution of a brain that tries to stay one step ahead of the world by constantly predicting what's going to happen next.

These three strategies—dissociation, compression, and prediction—form the holy trinity of personal narrative. Each process has evolved for different reasons and can be found in some form in almost every mammal. In humans they come together in a unique way that lets us construct narratives of our lives. Of course we use these stories to describe things that happen to other people, but the narrative we care the most about is our own. When we think about who we are, we tend to think about the narrative version of our life: the ongoing story of where we came from, our professions, spouses, kids, and so on. As we saw in the case of multiple personalities, the trinity can be deployed in different ways to create multiple narratives. Whatever you think of as your core personal narrative is only one of many possibilities.

To understand how these stories form in our heads, let's break down the narrative process.

On the surface, narrative feels like a form of magic. Magical thinking, at least. A sequence of events occurs. It doesn't matter if they are causally related because our brains will connect them to each other. Academics, like me, wag their fingers and intone that correlation is not the same as causation, holding up this type of cognitive error as the poster child of logical fallacies. We all know that just because two events happen in sequence doesn't mean the first caused the second.

But why, then, does it *feel* like they're connected? One possibility, which I just alluded to, is that we construct the illusion of causality. Narrative forms the glue linking together what would otherwise be a frighteningly random world.

Superstitions are classic examples of this type of narrative construction. Everyone has a few. For instance, I always wear a particular T-shirt when starting a new experiment. Why? Because when I do, the experiment seems to go well. I know it makes no sense, but the feeling is overpowering. Besides, what's the harm?

Baseball players are notoriously superstitious. Some of these superstitions are so common that they serve as unwritten rules of the game. Pitchers don't shave when they're starting. Don't dare step on the foul lines when going on or coming off the field. If someone is on a hot hitting streak, they stay with the same bat. Many will continue wearing the same underwear, too. And don't ever, ever talk about a no-hitter in progress for fear of jinxing it. Superstitions are powerful examples of narrative construction. They are minimalist, causally linking only two events, which makes them perfect building blocks for more complex narratives.

My T-shirt superstition is an easy example to break down. I know exactly how it came to be. In 2014, I was two years into a new project where I was teaching dogs to go in the MRI scanner awake and unrestrained. The goal was to decode what dogs think.[1] Knowing how much I loved the dogs in the project, my children bought me a special T-shirt for Father's Day. An internet meme had grown around this "three wolf moon" shirt. People started posting bogus Amazon reviews about how great things happened to them after wearing the shirt. I started wearing the shirt on scan days, just as a joke. Plus, it was a cool shirt. But dog fMRI is a tricky business. The dogs are all well trained, but sometimes they don't feel like participating. They get antsy in the scanner. After I wore the shirt a few times, it seemed like we had an uncommonly good string of scans. Maybe, I thought, there *was* something to the shirt. And that's how a narrative formed.

It is important to realize that this is not exactly a case of confusing correlation with causality. Causality means that something effects a change in something else. Because time is unidirectional, the causative event must temporally precede the result. I put the T-shirt on before the scan, so it meets the temporal requirement of causation, but that's all. Correlation means that two events tend to occur together. Correlation doesn't really come into play here, because if I always wear the shirt I never have the opportunity to test the counterfactual possibility that good things also happen when I don't wear the shirt. The result is that the shirt and the outcome become yoked together in my mind.

The problem is that correlation can be measured accurately only through random sampling. To test my superstition, I would have to flip a coin every experiment day and wear the T-shirt only if the coin comes up heads. Then I would have to make note of how the experiment goes on days that I wore the shirt compared to the days I didn't. Yeah, no thanks. Better to stick with a good thing.

Without consciously acknowledging it, I had been employing a common mental heuristic called *win-stay/lose-shift*. With this strategy, a person stays with whatever they're doing as long as the outcomes are good. Only when something bad happens do they switch to something else. This strategy appears to be at the root of most sports superstitions. Baseball players are likely prone to these beliefs because of the length of their season and the repetitive nature of the sport. Baseball has ample opportunity for long strings of hits, outs, wins, and losses (unlike, say, football). By chance alone, any player is bound to encounter hot and cold streaks.

As I've described it, win-stay/lose-shift sounds like a convoluted post hoc explanation for why humans do what they do, but animals do it, too. In fact, this strategy is a direct result of basic learning mechanisms that have been known for over a century. Edward Thorndike, a psychologist at Columbia University in the late nineteenth and early twentieth centuries, is generally credited with describing this foundational principle of learning. As Thorndike observed, "Responses that produce a satisfying effect in a particular situation become more likely to occur again in that situation, and responses that produce a discomforting effect become less likely to occur again in that situation." Thorndike's *law of effect*, as it is called, formed the basis for what behaviorist B. F. Skinner later called operant conditioning. Initially, neither Thorndike nor Skinner were referring to humans. Their conclusions were based on animal behavior.

While studying with William James at Harvard, Thorndike developed a puzzle box for cats, which he kept in James's basement.[2] The box was a simple crate with a trap door that the cat could open by stepping on it. Thorndike placed a cat in the crate and a saucer of milk outside. At first, the cat frantically tried to escape until it accidentally triggered the door to open. When Thorndike repeated the experiment, the cat would waste less and less time thrashing about, until finally, it would go straight to the door.

Psychologists have compared the law of effect to Darwin's theory of natural selection in terms of its explanatory power. The law of effect says that successful behaviors, much like successful organisms, will survive, while unsuccessful behaviors won't. It is survival of the fittest behavior. Rats, cats, dogs, pigeons—you name it—every animal that has been studied seems to follow the law of effect. So it should be no surprise that humans do, too.[3] The only difference seems to be that humans add post hoc explanations—narratives.

The law of effect provides a simple, yet powerful, explanation for the habits that we and other animals form. Although Thorndike didn't explicitly articulate it, an animal's action and the desired result must occur closely in both space and time. This seems to be true of superstitions as well. For example, I could have formed a superstitious belief between a successful experiment and what I ate for dinner the night before, but the shirt was in closer proximity. In space, I was wearing it at the MRI facility, and in time, I was wearing it simultaneously with the good outcome. Indeed, it is a basic principle of operant conditioning that learning occurs best when the outcome follows immediately from an action. Similarly, superstitious beliefs are more likely to form when events occur close together.

If the law of effect underlies superstitious beliefs, and ultimately our fantastical explanations for them, is it possible that animals can also be superstitious? Remarkably, the answer seems to be yes.

Among B. F. Skinner's many contributions to psychology was a paper titled, "'Superstition' in the Pigeon."[4] In this study, Skinner placed a hungry pigeon in a cage, much like Thorndike had done with cats fifty years earlier. A solenoid valve was rigged up to release food pellets every fifteen seconds. Initially, the pigeons moved about the cage, pecking at this and that. In accordance with the law of effect, whatever the pigeon happened to be doing at the time the food was released got reinforced. Unlike Thorndike's experiment, the food release was not contingent on anything the pigeon did. And yet, the pigeons quickly developed idiosyncratic rituals. One bird was compelled to walk in counterclockwise circles between reinforcements. Another repeatedly thrust its head into one of the upper corners of the cage. A third resorted to tossing its head in the air. Others swayed back and forth. Skinner thought these behaviors demonstrated a rudimentary form of superstition. He writes, "The bird behaves as if there were a causal relation between its behavior and the presentation of food, although such a relation is lacking."

Although Skinner was not normally concerned with what the pigeons were thinking, he did note the similarity to many human behaviors. It wasn't until the 1980s, though, that analogous experiments were performed in people. Gregory Wagner and Edward Morris, psychologists at the University of Kansas, conducted a simple replication of Skinner's experiment, but instead of pigeons, they used children.[5] Instead of food pellets, a mechanical clown named "Bobo" dispensed marbles at fixed intervals of either fifteen or thirty seconds. The children, who ranged in age from six to eight years old, were first familiarized with the test room. After they were comfortable with the setup, the researchers left them alone with the clown in sessions lasting eight to ten minutes and filmed what happened. Of the twelve children in the study, seven developed superstitious behaviors. Most of these were self-directed, like puckering the lips, thumb-sucking, or hip-swinging. But a few directed their behaviors at Bobo, like touching or kissing him. That same year, Koichi Ono, at Komazawa University, reported on a similar experiment in twenty Japanese college students.[6] Instead of a clown, Ono set up a booth with three levers. The levers did nothing. The students received points on a fixed schedule every thirty or sixty seconds. Three students developed a stereotyped ritual of manipulating the levers in a series of short pulls, followed by a long hold. Although only a few students evidenced persistent behavioral patterns, many others developed transient rituals.

These experiments, stripped down to the bare minimum of stimuli, like levers that do nothing and a clown that spits marbles, demonstrate how little it takes to form contingent responses, even when the contingencies are self-generated. The fact that these rituals are self-generated is the critical factor. Humans, like cats and pigeons, attribute outsize importance to their own actions. It isn't just the coincidence of two events happening in close proximity to each other that causes superstitions to develop. It is the proximity of an event to an act under the individual's control. The feeling of "I did that" leads to the illusion of control and ultimately superstitious behavior. It is an example of how we substitute movement in space to explain things that are merely sequenced in time.

All of these experiments targeted the development of superstitious behaviors rather than beliefs. Thorndike and Skinner, textbook behaviorists, were only concerned with what they could measure objectively. But humans aren't automatons. We think about everything. It should be no surprise that we attribute the same magical powers to our thoughts as to our actions.

NOWHERE IS THE POWER OF SUPERSTITIOUS BELIEF MORE evident than in how we think about sickness.[7] Doctors have long known about the waxing and waning nature of most illnesses. In only select circumstances can medicine intervene to substantially alter the course of an illness. Antibiotics can kill most bacterial invaders. Vaccines can prevent serious infection before they cause much damage. Cancer treatments have improved to the point that many malignancies can be cured. Those are the successes. But often doctors work to keep the patient alive, hoping that they will recover on their own. Other times, without intending to, they administer treatments that make the patient worse. It is no wonder that patients and doctors attribute magical healing powers to the strangest things. Like prayer.

Some people may consider praying a behavioral act. Depending on how it is done, it may exhibit some behavioral manifestations, like kneeling or pressing one's hands together, but most people do not consider these acts central to (or sufficient for) prayer. The key feature is an internal dialogue with God. Yet consider the timing. Sick people are most likely to pray when they are feeling their worst. Assuming they don't die, it is precisely at that point in time that they're likely to begin feeling better. In those circumstances, how could you not conclude that God heard your prayers and intervened? It makes for a perfect narrative. After all, most people don't pray for health when they're healthy. It is easy to see how a simple temporal coincidence between prayer and sickness can create a full-blown resurgence of faith.

Consider the story of Sister Boniface Dyrda. In the fall of 1959, Sister Boniface lay in her hospital bed staring at the ceiling.[8] Her hands were swollen, again. She could tell by the loss of sensation in her fingers. She lacked the strength to raise her arms, but she didn't need to see them to know they would resemble a string of kielbasa in the butcher's shop. Sister Boniface began to weep— not so much because of her overall weakness but because the mysterious illness had rendered her hands unable to manipulate the rosary beads that had given her so much comfort through her life. She began the recitation of prayers, but without the beads to mark her place, she quickly lost track of the Our Fathers.

She could hear the group of doctors coming. She heard the chief resident recite her case in the telegraphic code the doctors used: forty-three-year-old single white female, presents with relapsing fevers, abdominal pain, macular rash, and intermittent weakness and paresthesia of the extremities.

The pack of doctors debated among themselves before shuffling en masse into her room. Sister Boniface rolled on her back and saw the doctors forming a semicircle at the foot of the bed. She hoped that maybe they had some answers. The attending explained how a doctor in New York had discovered that some patients' bodies make proteins that attack certain organs. The skin and joints were usually affected, he said, but it could happen to the heart, the kidney, the nervous system. Anywhere, really. The disease was called lupus. In severe cases, the spleen, which normally filters the blood for infections, attacks the body's own blood cells. To attempt a cure, the doctors were going to remove her spleen.

The next two weeks blurred together in a fog of pain and morphine. Before she had fallen ill, Sister Boniface had weighed a healthy 140 pounds. By the time the surgeons removed her sutures, she was down to 86. The last time she had weighed so little was when she had joined the order. That was way back in 1929, when she was thirteen years old and everyone still called her by her birthname, Therese.[9]

Father Marion Habig, the order's chaplain, visited her frequently. He had already administered last rites. Without anything else to do, he asked her whom she had been praying to. Saint Jude, she said. Saint Frances Cabrini.

Habig told Sister Boniface she was praying to the wrong saints. They didn't need her prayers because they were up in Heaven already. Habig suggested Boniface pray to someone else. Habig was fond of an up-and-coming potential saint named Junipero Serra.[10] Father Habig thought they could get Serra beatified and urged Boniface to pray to him for a miracle. Sister Boniface had never heard of Father Serra, but she immediately latched onto the idea that Serra was looking for a miracle to work on. She could be that miracle.

In reality, it was Father Habig who was hoping for a miracle. In California, he had worked with another priest, Father Noel

Maholy, at the seminary of the Old Mission Santa Barbara who had inherited Serra's cause for sainthood. Serra had been considered for possible beatification and canonization as far back as 1934, and Mission Santa Barbara had been the headquarters of the cause. Maholy's predecessor had spent decades traveling between California, Mexico, and Spain, collecting the necessary information to make a case to the Vatican.[11] He had even survived an airplane crash in Ireland, but in 1949, while visiting New York, he was seriously injured in an automobile accident.

Although the priests had accumulated thousands of pages of documentation, they still lacked the most important piece of evidence necessary for sainthood: a miracle.

A week after Sister Boniface had begun praying to Serra, she sat up in bed and told her nurse she wanted to go for a walk. Later, she felt a quivering inside that she almost didn't recognize. She realized that, for the first time in months, she was hungry. She wanted an apple. Two weeks later, she was discharged from the hospital to finish recuperating at her convent. By the spring, she was back to teaching. Sister Boniface had gone from death's door to full recovery in a matter of months.

A divergence of conflicting narratives ensued. Her doctors knew well that lupus was a chronic illness of the waxing and waning variety. Symptoms came and went. The surgeons pointed to the removal of her spleen and took credit for the cure. After all, Sister Boniface had been getting worse before the procedure, and then, afterward, she got better. Sister Boniface had her own explanation—the narrative implanted by Father Habig. There was no doubt in her mind that Junipero Serra had answered her prayers.

Here was a conundrum. Everyone agreed on the basic sequence: Sister Boniface fell ill, possibly from lupus; she received several medical and surgical interventions and, despite these treatments, was close to dying; she prayed to Junipero Serra and within a few weeks started getting better. Given the sequence of events, what was the correct causal explanation? Spontaneous recovery? Medical cure? Or miracle?

Science and religion have always offered competing narratives, and in the modern era, science has usually won out. But news of Sister Boniface's miraculous recovery began to spread. It didn't take long for word to reach the priests in California. Father Maholy was skeptical. He had had his hopes raised before with purported miracles. The more he learned about Sister Boniface's recovery, though, the more inexplicable it seemed. And that, Maholy knew, was the key to Serra's cause.

It is a common misconception that the church needs proof of a miracle for sainthood. That is not quite right. All Maholy needed to do was build a case that Sister Boniface's recovery wasn't scientifically explainable. Although it had become harder and harder to find miracles in the modern era, Maholy had been on Serra's cause for twenty years. What was another twenty? A blink of an eye in church time.

The investigation into Sister Boniface's recovery was finally closed in 1987—nearly thirty years after the onset of her illness. Despite thousands of pages of medical records and testimony from medical experts, no one could offer an explanation for why she got better. The Vatican, therefore, concluded that Sister Boniface had been cured through "Serra's miraculous intercession."[12] On September 25, 1988, Pope John Paul II beatified Serra in a ceremony in Rome attended by Sister Boniface herself.

Although everyone assumed that Sister Boniface had had lupus, it was never medically proven. Diagnosing lupus is not as straightforward as, say, diagnosing diabetes. In the 1950s, the diagnosis was based on symptoms alone. Today, doctors look for the presence of autoimmune antibodies targeting a range of proteins in the body. Even then, diagnosis can be difficult. It is possible she never had lupus.

The tale of Sister Boniface and Junipero Serra is an example of a classic narrative trope: the hero's journey. Everyone could agree on the temporal coincidence of Sister Boniface's improvement following her prayers to Serra, but in the telling, there emerged two distinct interpretations. For the faithful, Sister Boniface's narrative transformed from that of a sickly nun wasting away in a St. Louis hospital to one of a woman who launched a California priest, dead for two centuries, toward sainthood. For the skeptics, this was a narrative of how modern medical science saved a dying woman. This schism demonstrates the power of belief in shifting narratives.

As we're discovering, not all narratives are created equal, at least to our minds. We irresistibly favor some and discount others. In the next chapter, we will examine the most common narrative forms and why they are so prevalent in the stories we tell ourselves.

CHAPTER 8

Narrative Forms

I N THEORY, THERE EXIST AN INFINITE NUMBER OF STORIES, and infinite ways to tell them. In practice, though, we keep coming back to the same few. These are the stories that stick with us throughout life. As we'll see in this chapter, certain narrative forms have evolved to gain more traction in our brains than others, with one in particular developing outsized importance over the others.

Even though writing is a recent invention, we have good reason to believe that the telling of stories is much older than that, maybe as old as humans themselves. Without a written record, we can never know what ancient humans said to each other, but there is plenty of archaeological evidence that indicates our ancestors were intensely invested in sharing significant events with each other. One of the most famous examples is the Lascaux caves in France, which contain several hundred murals. The images are believed to be at least seventeen thousand years old, predating the invention of writing by twelve thousand years. The paintings contain depictions of animals and of humans hunting them. Looking at them, one has the overwhelming sense that the murals convey momentous events. We can imagine these hunters gathered around a fire, recounting tales of a great hunt.

There is another striking aspect of the Lascaux art that tells us that these images represent stories. They are dynamic. These are not pictures of animals standing around. The animals are captured in motion. Herds of wild horses are in full gallop, hunters flinging spears. Thousands of years later, we can easily get the gist of what happened. Pretty amazing, actually. This dynamism gets to the heart of why we tell stories. When something in the world moves, it tells us that something has changed. As Isaac Newton wrote, movement is evidence that some force has caused a ripple in the world. A gunshot rings out and someone collapses, the world around him changed forever. These are singular events, marked by the scale of the change. Of course, there are causal explanations, and the sequence of events that led up to them gives us clues. But we do not remember the sequence exactly. We remember a highly edited, curated version in the form of a story.

As an experiment, try to recall the events of 9/11, or some equally momentous event, in reverse. It's hard, isn't it? If our memory was organized as a simple sequence of events, like a digital recording, then you could play them back in reverse order. Our inability to do that demonstrates that a structural form has been imposed on our memory. Our brains organize information in a way that makes sense to us. Most of the time, that organization takes the form of a story.

I would go further, though, and suggest that storytelling is so deeply intertwined with the biology of the human brain that it isn't simply that we use stories to understand the world around us. Stories dictate our perception of reality, imposing temporal order, and we live the delusion that each of us is the protagonist of these stories.

WHEN I WAS A CHILD, MY PARENTS KEPT A HARDCOVER TOME of stories in the living room. In my mind's eye, I remember it as a teal, perhaps greenish, book the same size as the encyclopedia volumes that it sat next to. At three or four years old, it took me both hands to dislodge it from its location and carry it over to the couch for story time. The book, now sadly out of print, was called *The Golden Treasury of Children's Literature*, edited by Bryna Untermeyer and Louis Untermeyer. Within this precious volume, the Untermeyers had collected seventy-one excerpts of the best-known fairy tales and children's books. It included everything from Aesop's fables to Rudyard Kipling, the Grimm Brothers, Hans Christian Andersen, Uncle Remus, and Ray Bradbury. Beautifully illustrated, it was my favorite book.

Although I could choose from pleasant fare like Winnie-the-Pooh or Alice in Wonderland, I invariably begged for the story of Bluebeard. The version in the book was adapted from the seventeenth-century French folktale by Charles Perrault. In the story, Bluebeard, a rich nobleman, seeks a new wife. It is well known that his previous wives have disappeared. Yet his neighbor marries his youngest daughter off to Bluebeard anyway, and Bluebeard ferries her away to his castle. Soon, Bluebeard has to leave on business, but before he goes, he gives his wife the keys to the castle. He warns her she can go anywhere except the room in the tower. You can guess what happens next. Unable to resist her curiosity, the youngest daughter eventually unlocks the door to the forbidden room, where she discovers Bluebeard's previous wives hanging from meat hooks. She drops the key in terror, but it's a magical key, and no matter how hard she tries, the blood from the floor will not wash off. Meanwhile, Bluebeard returns unexpectedly. As soon as he sees the bloody key, he knows what has happened and says that his wife must die for disobeying him. She convinces him to wait until the next morning, giving her time to enlist her visiting sister to fetch help. The sister returns with their brothers just in time to kill Bluebeard. The neighbor's daughter inherits his fortune and eventually remarries, living happily ever after.

Bluebeard has all the basic elements of a classic fairy tale: a journey away from home, a villain, peril, a magic key, and the ultimate vanquishing of the villain. It's a grisly tale, and I'm sure the gore factor was a major reason for its persistence in my memory.

Fairy tales are more than simple stories to entertain children. They almost always have some deeper meaning than the story itself (you can read Bluebeard as a simple cautionary tale or as a metaphor for Eve and the consequences of sexual knowledge, among other interpretations). Moreover, a closer look at these tales reveals that almost all of them follow the same narrative arc: a hero's journey. In fact it is this arc that makes them fairy tales. Because fairy tales are some of the first stories that children hear, they become embedded in our brains in ways that we are scarcely aware of, forming the backbone of our own personal narratives.

WHAT IS IT ABOUT FAIRY TALES THAT THEY HAVE SUCH power over our imaginations? In addition to being some of our first stories, they're generally short, so they are easily absorbed in the limited attention span of their target audience. And because fairy tales all follow the same form, they benefit from repetition and the pleasure children gain from being able to recognize and anticipate what's coming next. This same quality also makes them easier to analyze structurally and allows us to detect common themes.

There exist several classification systems for fairy tales, but one created by a Russian folklorist in the 1920s has stood the test of time. Vladimir Propp was born in 1895 in St. Petersburg, where he also later taught German to high school and college students. His passion, though, was Russian folklore. In 1928, he published a landmark book called *Morphology of the Folktale*.[1] Propp analyzed the structure of one hundred Russian fairy tales and discovered that they all followed the same structure, which he divided into thirty-one discrete functions that followed the same sequence. These functions, he wrote, "serve as stable, constant elements in a tale, independent of how and by whom they are fulfilled." The first function, for example, is when a member of the family leaves home. This is followed by an interdiction in which something is forbidden. In the first *Star Wars* movie, a classic fairy tale if there ever was one, Uncle Owen forbids Luke Skywalker from searching for his father, telling him that Anakin Skywalker is dead. In Propp's model, that function is always followed by the hero's violation of the interdiction. Next a villain enters the picture and gains information—as during the interrogation of Princess Leia by Darth Vader and Captain Tarkin. This leads to the unwitting aid of the villain, as when Leia admits to them that the rebel base is on Tatooine (Tarkin blows up her home planet of Alderaan anyway).

Amazingly, Propp described all of these elements, including the harm to the victim's family. And these are just the first eight functions. According to Propp's schema, the hero then leaves home (bar scene at Mos Eisley), is tested (Luke learns to use a lightsaber), acquires the use of a magical agent (the Force), and joins the villain in combat (Luke and Han Solo rescue Leia and battle to escape the Death Star). After this, the initial misfortune is resolved (R2D2 delivers plans of the Death Star to the Rebellion). The hero returns but is pursued (TIE fighters chase the Millennium Falcon). A false hero presents unfounded claims (Han Solo claims the reward for rescuing Princess Leia). A difficult task is then proposed to the hero, which he then resolves (Luke blows up the Death Star). The hero is recognized and the villain punished (Vader spins off into space while Luke is hailed as the hero). Finally, the hero is married and ascends the throne—cue the ceremony at the end of *A New Hope*.

The basic narrative of *Star Wars*, indeed of all fairy tales, pre-dates even Propp's analysis. Joseph Campbell, the famed professor of literature, has argued that all stories have their origin in a sin-gular form. In 1949, Campbell published *The Hero with a Thousand Faces*, which would become perhaps the most influential book on the universality of myths.[2] By analyzing not only folktales but also the myths and stories of many diverse cultures, Campbell made the case that the hero's journey, or what he called the *monomyth*, is not only the most popular story of all time but also the only story.

Campbell simplified Propp's structure by dividing the journey into three major sections, which is reflected in the modern story-telling trope of three acts. In the first act, which Campbell calls *Separation*, the hero (Luke Skywalker) encounters a call to adven-ture, which at first he refuses. This is followed by some form of supernatural aid, which causes the hero to cross a threshold and enter "the belly of the whale," meaning some kind of dream land-scape. In the second act, or *Initiation*, the hero enters a road of tri-als. Here, he meets a goddess (Princess Leia) and is tempted (the Dark Side). He confronts a father figure (Darth Vader). There is an apotheosis in which someone dies and is deified (the death of Obi-Wan). In the third act, titled *Return*, the hero returns home a changed person. At first he refuses to return, but this is inevitably followed by a rescue (when Han Solo swoops in at the last minute, preventing Vader from attacking Luke, allowing Luke to complete his mission to destroy the Death Star). The hero then returns for real as the master of two worlds (Luke is honored in the closing ceremony).

If this all feels rather neatly mapped, you will not be surprised to learn that the three-act structure has been taught for decades at USC, where *Star Wars* creator George Lucas studied film. Lu-cas, well versed in classical mythology, has often spoken of Joseph Campbell's influence on him and, in fact, purposely set out to cre-ate a modern telling of the monomyth in the original *Star Wars*.

Star Wars has since become canonized as the archetype of a hero's journey and has inspired countless imitations. That's by design.

The ubiquity of the monomyth is not to be underestimated. Whether we're binge-watching television, reading a book, or even reading the news, nearly every story we encounter follows this form to a large degree. Our brains expect it. Stories that deviate from the form jar the senses. We can tolerate the occasional detour, but not on a regular basis. Inevitably, we find ourselves returning to the quaint notion that "life is a journey." The hero's journey becomes a guiding theme in who we think we are. And because this story gets embedded in our brains from an early age, its arc finds its way into everything we perceive and what we ultimately remember.

THE HERO'S JOURNEY IS NOT ONLY THE FIRST STORY THAT children hear, it is the first story of all humankind. The hero's journey is so old, it may be the first cultural artifact baked into the human brain. This is easy to verify because all one has to do is look at the first story for which we have a written record.

The Sumerian *Epic of Gilgamesh* tells the story of a man named Gilgamesh, who lived between 2850 BC and 2700 BC (though his age was probably exaggerated).[3] The Sumerians, who resided in northern Mesopotamia between the Tigris and Euphrates Rivers (now Iraq), in roughly 4000 to 2000 BC, are generally credited with the invention of writing. The writing was cuneiform, which appears as complex sequences of wedge-like marks pressed into clay tablets. It was much closer to an alphabet than the later Egyptian hieroglyphics. It wasn't until the 1800s that these tablets were deciphered after the discovery of an ancient monument on a cliff in western Persia (now Iran). This monument contained an inscription in three languages, including Akkadian, which is what the Sumerians spoke. Like the Rosetta Stone, the monument enabled the translation of many Sumerian tablets. Twelve of these tablets contained the *Epic of Gilgamesh*.

The epic is a perfect example of the hero's journey and can be read as a three-act adventure. In the first act, we are introduced to Gilgamesh, King of Uruk and the son of Gods. Gilgamesh, we learn, is revered beyond all previous kings. He possesses all wisdom, is the strongest man alive, and, as is the right of a king, sleeps with all young women before they are married. This causes the people great distress, and they complain to the Gods. In response the Gods create Enkidu as Gilgamesh's equal. Enkidu is born of the wilderness and lives naked with the animals. Word reaches Gilgamesh about this terrifying wild man. Gilgamesh sends Shamhat, the temple prostitute, to seduce Enkidu, and after seven days of lusty enjoinder, Enkidu's animal friends abandon him, and he is forced to leave the wild. On their way back to Uruk, Enkidu meets a shepherd who tells him what Gilgamesh did to their daughters. Enkidu, outraged, challenges Gilgamesh. But Gilgamesh is too strong, and Enkidu concedes defeat. After this, the two become best friends.

In the second act, Gilgamesh decides to take Enkidu with him to slay Humbaba, the monster that guards the Cedar Forest. Humbaba tries to scare them off by bellowing from the mountain and unleashing earthquakes. When they engage in battle, Gilgamesh calls for his God friends to help him. Shamash, the Sun God, and the most powerful of the Gods, sends all the winds at his disposal, including tornadoes, typhoons, and dust storms. It is too much for Humbaba, and Gilgamesh kills him. After seeing what Gilgamesh can do, Ishtar, the goddess of love, wants to marry him and have his children. Gilgamesh declines her entreaties. Ishtar, beside herself, curses Gilgamesh, but her curse misses Gilgamesh and lands instead upon his friend. Enkidu develops a fever and, after twelve days, dies.

In the third act, Gilgamesh grieves for his friend. And now aware of the possibility of his own death, he seeks immortality. He sets out on a journey to find Utnapishtim, the only human to have survived the Great Flood and to whom the Gods have granted immortality. He meets Shamash, the Sun God, who cannot understand why Gilgamesh continues to search for something that he will never find. It is here that we arrive at the heart of every hero's journey: the human confrontation with mortality.

The *Epic of Gilgamesh* contains obvious parallels to both the *Odyssey* and the Book of Genesis, reflecting their common origin in the stories being told at the beginning of human civilization. Although Gilgamesh is the oldest written story, oral versions of Gilgamesh, the *Odyssey*, and Genesis are believed to have been told for millennia, perhaps since the beginning of human cities.

IT IS SAFE TO SAY THAT THE HERO'S JOURNEY HAS BEEN PART of humanity at least since Gilgamesh, about five thousand years ago, but probably much longer. Collectively over millennia, and individually over our lifetimes, we have been conditioned to respond to the form of the journey. Our brains cannot help but interpret a story in terms of this narrative arc. There must be a protagonist. An external event upsets their life, and this spurs them to action. They pursue their desire for something, while encountering various struggles. Eventually, they make a choice, the outcome of which changes them forever.

Of course, the hero's journey isn't the only narrative scheme collectively stuck in our brains. For example, there is boy-meets-girl and all its gendered (or not) variations, the rags-to-riches story, and many others. How many? Literary scholars debate this question endlessly, but in 2017, researchers at the University of Vermont analyzed the shape of story arcs of 1,327 books of fiction archived in Project Gutenberg and concluded that there are only six story forms.[4] The first, rags to riches, is central to the mythology of the American Dream. Under this category fall Shakespeare's *The Winter's Tale* and *As You Like It*, Jack London's *The Call of the Wild*, Rudyard Kipling's *The Jungle Book*, and Oscar Wilde's *The Importance of Being Earnest*. There's also the opposite form, riches to rags, exemplified by *Romeo and Juliet*, *Hamlet*, *King Lear*, and H. G. Wells's *The Time Machine*. These stories are classically considered tragedies, where the hero is doomed from the start. In the third form, "man in a hole," the protagonist starts on a high note, gets in a jam, and then climbs out. Well-known examples include *The Wizard of Oz*, *The Prince and the Pauper*, and *The Hobbit*. The fourth form, "Icarus," is marked by a rise and then a fall and includes *A Christmas Carol* and *Paradise Lost*. The hero's journey belongs to the fifth form, aka "Cinderella" (rise-fall-rise). These include *Treasure Island* and *The Merchant of Venice*. Finally, the sixth form is named for *Oedipus* and charts a fall, followed by a rise, and then a fall.

When these six forms are plotted in a graph, the differing trajectories become apparent, revealing the basis functions for most of the stories that you have consumed throughout your life. Although I have suggested that the hero's journey is the predominant form in our personal narrative, it is possible, even likely, that we deploy the others when adopting other identities, as we saw in the last chapter. Alter egos, like Mr. Hyde or Eve Black, because they are the opposite of the prime personality, have to follow a different arc. If the prime is following a Cinderella arc, then their alter is likely following the Oedipus arc.

Although you are probably familiar with these stories, you've likely forgotten some of the plot details. But it doesn't really matter. One of the remarkable features of stories is how they roll up so much information into a compact form. Great stories are so efficient in this regard that the protagonist can embody the entire

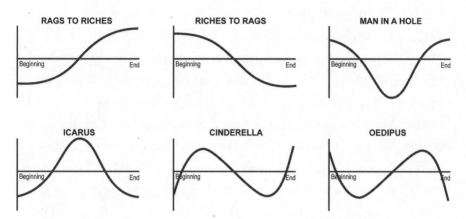

The six canonical story arcs. *Reagan et al., "The Emotional Arcs of Stories," 1–12.*

story. You may not remember the exact plot of *The Great Gatsby*, but who can forget Jay Gatsby and his representation of the hollow aspirations of the American Dream? Or Daisy Buchanan, her beauty and her careless social climbing? One might say they are emblematic, standing as symbols for the author's themes, but that's too simplistic, flattening unforgettable, complicated characters. I think it goes deeper than symbolism. They are compressed representations of sophisticated social struggles. They are coiled up in our memory banks, ready to be unwound when the circumstances call for it.

THE CRITICAL QUESTION, THEN, IS HOW DOES A CHARACTER become the dominant template for one's own life narrative? Some may argue that fictional characters serve no deeper purpose than entertainment; however, as we've seen, it's undeniable that a well-wrought protagonist can make an especially powerful impression on a developing mind.

At some point, everyone has encountered a story that resonates so much with them that they say to themselves, "That person is just like me." Or, whether consciously or not: "I want to be like them." Think of the books you read during those developmental years and the characters whose shoes felt most comfortable to you. Harry Potter. Hermione Granger. Paul Atreides. Katniss Everdeen. Elizabeth Bennet. Charlie Bucket (preteen, but still, who doesn't want to own a chocolate factory?). Holden Caulfield. Ignatius J. Reilly? With the exception of the last two, all are heroic, overcoming obstacles on their journeys to becoming better people.

The formative years of adolescence and young adulthood are a critical period for figuring out what type of journey one is on. Stories like that of Jay Gatsby or Anakin Skywalker, although meant to be cautionary tales, also have the potential to be incorporated into a person's self-identity during this period. Although I do not think this is a common outcome of adolescence, the assumption of the tragic character as a life theme occurs with enough frequency that it bears mentioning. To be clear, I do not view this as a healthy direction, no matter what adversity a person has faced. Such people may wear their victimhood on their sleeve.[5] And although one might think that suffering trauma (real or perceived) would make a person more empathetic toward others' misfortunes, often it is the opposite, where one's sense of entitlement grows so large that the person becomes numb to others' suffering. Victimhood may also lead to excessive rumination as a person dwells in the past.

In Jay Asher's *Thirteen Reasons Why*, the victimhood narrative reaches a performative apex. In this hugely popular 2007 young adult novel, we meet Hannah Baker, a teenager who has committed suicide. Her story is told through thirteen sides of cassette tapes she recorded before she died. The box of tapes is sent to her friend, Clay Jensen, with the instructions to listen to the tapes and then pass them on to the next person. She prefaces the instructions by warning the recipients that she has made copies of the tapes and arrangements to release them in a "very public manner" if the box doesn't reach the final person, her teacher, Mr. Porter. Flipping back and forth between Clay's and Hannah's narratives,

Asher paints a portrait of a painful high school life. Hannah went through the usual adolescent struggles with boys and mean girls and rumors and bullying. There are episodes of sexual assault. Hannah becomes depressed and begins to formulate a plan to kill herself. In the final tape, she hints at these feelings to Mr. Porter, who is clearly concerned. But Hannah blames him for not caring enough to stop her.

Hannah is doomed from the start. She's dead before the narrative even begins. There is no possibility of a hero's triumphant return. From my adult perspective, I see a depressed, self-absorbed girl struggling with issues that many teens do. Certainly no one would advocate suicide as a solution. Even the manner in which the story is told paints a picture of a vengeful person. Why record all these tapes if not to inflict pain on the recipients?

An adult reading this book—or watching the Netflix TV adaptation—is likely to have a very different reaction from a teenager's. Indeed, for several years after the book's publication and then again after the Netflix release in 2017, researchers and mental health advocates worried about the uptick in teenage suicide, notably after the streaming version.[6] Other researchers disputed the correlation, finding no evidence for a causal relationship.[7] But even if a spike in teenage suicide cannot be attributed to *Thirteen Reasons Why*, it is easy to appreciate how Hannah's character might resonate with teenagers. Adolescence is already an age of self-absorption. A superficial reading of *Thirteen Reasons Why* may have an outsized influence on teens who find themselves in similar circumstances. For them, there is a real danger of incorporating Hannah into their own personal narrative. This does not mean books about teenage angst should be banned. Quite the opposite. They offer opportunities to engage with adolescents about the issues that are important to them while offering counterpoints to some of the potentially negative narratives they might contain.

THESE NARRATIVE ARCS, WHETHER THE HERO'S JOURNEY OR one of the other five forms, constitute the set of basis functions that allow us to connect events in time. And even though we may not consciously think of them every moment of the day, these ready-made narrative arcs serve as templates connecting past-, present-, and future-selves. In other words, the story of who we think we are, which we're taught to believe is unique in all of human history, is really a riff on a few very common themes.

By now, it should be clear that the notion of self-identity is a sort of delusion. Who you think you are is a construction. You are physically different from the person you were ten years ago, and even your brain has changed to the point that one cannot be sure that the physical substrate that holds the memories of your former self has any ground truth. Ample research has shown the fallibility of human memory. We are left with the unsettling prospect that the sense that we are the same person as these past versions is itself a fictional narrative. We have also seen how every experience is colored by one's expectations, which in turn are formed by past memories and the narratives we construct to explain why things happen the way they do.

If the stories we consume shape our sense of self, how then are we to draw boundaries between our own experiences and those that we read or hear about? The distinction for teens may be especially fuzzy, in part, because they haven't had many experiences to call their own. But even as adults, the boundary between self and other turns out to be not as firm as we might think. In the next section, we'll see how we get into each other's heads and how this blurs the lines between what are actually one's own thoughts and others'.

PART II

THE SELF AND OTHERS

CHAPTER 9

Flavors of You

In Part I, we focused on the individual construction of self, on how the notion of who we are is a confabulation that stems from the existential need to connect versions of ourselves across time. But the self is not just spread out temporally. It also expands and contracts in response to social demands. Humans are a social species, and we all have different versions of ourselves that emerge at work or at home or with friends and family. These facets of personality are more than a simple revelation of different sides of a person. I previously mentioned cosplay, raising the question of which is the true version of a person: the cosplayer or the version at work? We have already seen how dissociation is a normal part of the human experience, so it is not farfetched to believe that social context dictates who you think you are, perhaps to the point of your becoming a different person as a situation demands. The singular self, then, is a delusion not only across time but also across environments, allowing you to believe that you are the same person when you're with others. Examining the cracks in this delusion can shine some light on the mutability of self.

If the self is more fluid than we realize, then this fluidity is a direct result of long-standing evolutionary pressure to get along with each other. In Part II, we will look more closely at how the human constraint of living in a society results in the sharing of ideas. Much like what happens when we absorb stories from books or movies, this direct free flow of ideas among people further blurs the line between thoughts that originated in our own mind and ideas that we've assimilated from others. The provenance of one's thoughts gets muddied rather quickly in a uniquely human confusion.

Many animals are highly social, but only humans have evolved the ability to think about other people's thinking, and it is this skill that so readily lets us incorporate other people's ideas into our own. Researchers call this skill *theory of mind*, or ToM for short. As we'll see, ToM is a powerful evolutionary adaptation because it allows people to collaborate. A side effect, though, is the co-optation of individual thoughts into a collective soup that gets redistributed to the other members of a group. You might say that our brains evolved to readily accept the opinions of others.

EVOLUTION EXCELS AT SELECTING OPTIMAL STRATEGIES. THE law of large numbers says that a group is more likely to render an accurate judgment than any individual. In most situations, therefore, you are probably better off going along with the crowd than trying to go your own way. If following the herd, on average, is an optimal course of action—meaning that no other strategy can do better—then this impulse has been deeply embedded in the functioning of our brains by the forces of evolution.

Staunch individualists may be troubled by this thought. For my part, I find it helpful to consider the human tendency toward collective thinking as a feature, not a bug. It's analogous to the advice financial advisers give to invest in index funds rather than actively managed mutual funds. It's been shown repeatedly that high-flying fund managers who achieve stellar returns eventually regress to the mean. Over the long haul, it becomes increasingly difficult to obtain better returns than simply investing in the market as a whole. Our brains instinctively know this. But the individualist narrative that we create in our heads gets in our way. The hero must leave the comfort of the crowd to go on their journey, forge their own path, and slay monsters, before eventually returning home and bestowing their wisdom onto the community.

This is an aspirational myth. A useful one, to be sure, but still a myth.

There is another reason our brains have evolved to be more alike than different. Being of like mind makes it easier to cooperate. The ability to see things from another person's perspective forms the backbone of a society. The skill is so important that the inability, or unwillingness, to see others' points of view destabilizes societies.

In his 1755 *Discourse on Inequality*, the Swiss philosopher Jean-Jacques Rousseau famously addressed the problem of the individual's place in society. Describing the natural state of man as similar to that of other animals, that is, concerned entirely with one's own self-preservation (*amour de soi même*—love of self), Rousseau depicted the natural man as a savage. Complications arise, he showed, when natural man encounters others. At some point, natural man extends his self-interest to the environment in which he lives. He stakes out a plot of land and says, "This is mine." Rousseau claimed this single act represented the beginning of society and the civilization of humankind. It codified class inequality between the rich (those who owned land) and the not-rich. In Rousseau's view, the segregation into haves and have-nots created an obligation on the part of those with means to create a society that allowed for the freedom of those who were not well off. He

called this the social contract. Rousseau's writings came at a pivotal moment in the Enlightenment. Along with John Locke's and
Thomas Hobbes's, Rousseau's writings provided an intellectual rationale for the Declaration of Independence and the American
Revolution. In essence, King George had broken the social contract with the colonies. The only alternative was to form a new
government that reflected the will of the people.

As for microlevel social interactions, or how individuals be-
haved with respect to others, Rousseau had plenty to say about
that, too. The so-called stag hunt problem neatly encapsulates the
tension between self-interest and the common good. Although
primitive man was concerned primarily with his own well-being,
and therefore viewed other people with indifference, eventually
he came to realize that other people tended to behave in a similar manner toward him. From this, he reasoned that they probably thought as he did, and that it may be mutually advantageous
to adhere to a common set of rules to augment each individual's
well-being.[1] He learned when it was appropriate to rely on the
help of others and when a conflict of interests might cause trouble.
To illustrate this idea, Rousseau offered a simple parable of a stag
hunt. A deer provides more meat than any individual could consume by himself. But deer are notoriously wary creatures, excelling
at evading hunters. It was far more efficient to hunt as a team. With
two men taking up separate posts, the chances of success increased
dramatically. Any kill could then be shared. As Rousseau wrote,
"Every one saw that, in order to succeed, he must abide faithfully
by his post." But Rousseau was also aware of the more selfish side
of human nature. He wrote, "If a hare happened to come within
the reach of any one of them, it is not to be doubted that he would
pursue it without scruple." Having bagged a rabbit, he would,
Rousseau believed, abandon his post and his companions.

Rousseau's example is a general one: he never specifies the relative probabilities of success for stags and hares, nor the number

of hunters involved. Key to the stag hunt story is that a share of deer meat is more valuable than what a single hare provides. While tempting and immediately gratifying, pursuing rabbits is in nobody's interests. The Scottish philosopher David Hume, a contemporary of Rousseau, made a similar case about a boat.[2] Two men are stuck together in a rowboat. If they both row, they will get where they want to go. If only one rows, they will go in a circle, getting nowhere. Moreover, the person who rows wastes energy, resulting in an outcome worse than if neither had rowed at all.

In the stag hunt, self-interest and the common good are in complete alignment. (By contrast, in the better-known prisoner's dilemma, another classic thought experiment, self-interest and the common good are in direct conflict.) What is best for the individual is also best for everyone. Yet people often choose not to cooperate in the stag hunt. The best course of action depends on what the other person does. Either both participants hunt stags, or they both hunt hares. In game theory, these are called *equilibria* because once two people have settled on either of these outcomes, an individual can only do worse by choosing to do the opposite thing. But reaching a mutual agreement to hunt stags is difficult. Even though everyone might agree that it's the best thing to do, each individual has to trust that the other will do the right thing. This uncertainty in knowing what the other person will actually do is called *strategic uncertainty*. The mistrust of each other leads individuals to forgo the best outcomes in exchange for a sure thing. (In this case, abandoning the deer to grab the hare.)

At this point, you may be wondering: What does all this have to do with the self delusion?

The stag hunt dilemma appears over and over again in modern life, and it gets to the heart of how we calculate risk, of whether we think differently (and thus can be considered different people) when in isolation or when we're thinking of others.

Any activity involving teamwork can be considered a stag hunt. Each individual must contribute to the common good in order to reap the rewards. When scaled up, the inability to coordinate leads to the "tragedy of the commons." The original example of the tragedy of the commons told the story of English cattle farmers allowing their herds to graze on common land, a traditional practice. If an individual farmer increased their herd size, they could make more money. But if everyone did this, the land would be depleted and everyone would suffer. Thus, the tragedy is the destruction of the common good in service of the individual. It is precisely this dynamic that is responsible for the greatest threat to humankind: climate change.

Game theory provides a way to mathematically analyze the stag hunt, giving us a sense of why players might choose one strategy over another. In its simplest form, the stag hunt is a game between two people. Each must commit to an action—hunting hares or stags—and the payoff is determined by their joint choices. It doesn't matter whether the players are allowed to communicate with each other, because people sometimes lie. What matters is what each person actually does. Game theory tells us that the two equilibria are when both players hunt stags or both hunt hares. Stag is the outcome with maximum payoff and is, therefore, called the payoff-dominant equilibrium. The hares outcome, though having a smaller payoff, minimizes the risk of coming up empty-handed and is called the risk-dominant equilibrium.

Where the players end up depends on two factors: the relative payoff of stags to hares and each individual's tolerance for uncertainty. If a stag is worth a lot more than a hare, then people will tolerate the uncertainty of what the other person will do. On the other hand, if a stag is worth only a little more than a hare, then it is not worth the risk.

Economists disagree on whether strategic uncertainty is a manifestation of a person's general attitude toward risk or arises from an individual's propensity to trust other people. A third possibility is that when we find ourselves in situations that require the coordination of actions with other people, we become a different person, altering our mind to think more like the other's. If we could examine what is happening in people's brains while they make social decisions, then we might have a better idea of which mechanism is more important for coordination and the degree to which an individual changes who they are to suit a particular situation.

IN 2012, MY LAB SET OUT TO LEARN MORE ABOUT HOW individuals' risk attitudes change when dealing with other people. Using fMRI, we wanted to use the stag hunt game to probe the difference between simple risk attitudes and the uncertainty associated with thinking about other people thinking. It seems obvious that if two people are to solve the stag hunt problem and avoid being trapped in the risk-dominant outcome, they have to think about each other. Sure, you can avoid thinking about the other person, but then you will be forever relegated to hunting hares.

Thinking about what someone else is thinking is called *mentalizing*. This is not to be confused with the entertainment art of mentalism, although mentalists are experts in reading people. Rather, mentalizing is a skill that we all possess, to varying degrees, that lets us see things from another person's perspective. In other words, ToM.[3] In order to mentalize, though, you have to have a model of how another person's mind works. Seeing as you only have access to your own mind, you have to assume that other people's minds operate in a fashion similar to your own.

Given the complexity of mentalizing, it's not surprising that multiple brain regions are involved in ToM. Although some researchers have argued for a specialized ToM network in the brain, I favor a modular view of these regions.[4] Most brain regions, especially cortical regions, are routinely repurposed for multiple cognitive functions. ToM is no different. In fact, ToM is a version of the prediction functions discussed in Part I. The only difference is that with ToM, you're trying to predict what another person is thinking. And since you can never truly know what another person is thinking, your theory of someone else's mind is a pure simulation. Like reading a novel, ToM puts you in the shoes of another person. Such a rich simulation must marshal many brain circuits. You need sensorimotor regions to simulate their physical sensations and ac-tions. You need regions like the amygdala and insula to simulate their internal states, feeling what they feel.

You also rely on brain regions to ascribe reasons for another person's actions. For example, if you see a masked man walking into a bank, what assumptions do you make about what he's there to do? In the pre-Covid era, you might have guessed that he was hiding his identity so he could rob the bank. After March 2020, you might just assume he's avoiding a deadly virus. That form of causal reasoning relies on simulating his potential states of mind. Which states of his mind come to the fore of your mind, though, will be dictated by your own experiences and biases—in other words, your personal narrative. That is how ToM works. You draw on your own mental states to simulate someone else's.

It seemed a reasonable hypothesis, then, that ToM would pro-vide a way out of getting trapped in the risk-dominant outcome of the stag hunt. After all, putting yourself in the other hunter's shoes can help you see the mutual benefit of cooperating. How-ever, it might not be the only way. It's possible that people playing the stag hunt simply view the other player as a random process. In this view, you wouldn't have to simulate their mental state. You could just assign probabilities that they will hunt stags or hares and then make your choice according to the relative payoffs and your tolerance for risk. In this worldview, other people are simply automata whose actions can be treated like rolls of dice.

Since my team couldn't tell which strategy a person was using just from their choices, we set out to see whether brain imaging could distinguish between the two possibilities. If someone was mentalizing, we'd expect to see activation in the brain regions associated with the functions I mentioned above. If, on the other hand, a person reduced the other player to a lottery, then we should see activation only in regions involved in risk and reward.

We set up a series of stag hunt–like games for a person to play while we scanned their brain.[5] In some games, they played against a person, just like in the real stag hunt. In other games, we replaced the other person with a lottery having the same payoff possibilities. If the activation patterns in these two conditions looked the same, that would build the case for a person converting the stag hunt to a simple game of chance. If the activation was different, then that would suggest other processes came online during the stag hunt.

The first thing we noticed was the effect on the number of payoff-dominant choices (stag). In the lottery games, participants chose the payoff-dominant outcome 43 percent of the time. In the stag hunts, though, this increased to 73 percent. Simply knowing you were playing against another person nearly doubled the rate of cooperative choices, suggesting that, as we had suspected, different cognitive processes were being invoked by the stag hunt.

The behavioral data clearly demonstrated that there was some type of social process involved in the stag hunt. There were two possible explanations for the findings. It might be that the core cognitive processes in the stag hunt and the lottery were essentially the same, but the presence of a person in the stag hunt simply shifted people to the payoff-dominant choice. First, we used the brain imaging data to see which circuits were involved in the payoff-dominant choice irrespective of whether it occurred in a stag hunt or a lottery. This analysis revealed a broad swath of activity extending from the occipital cortex in the back of the brain up over the top to the parietal lobe. These regions are known to be involved in visual attention. The simplest explanation was that opting for the payoff-dominant choice required a closer examination of the relative payoffs, and this was true regardless of whether you were playing a stag hunt against a person or an inanimate lottery.

We then focused our analysis on the payoff-dominant choices. In these choices, we found noteworthy differences between the stag hunt and the lottery. A region called the cuneus was significantly more active in the stag hunt. The cuneus is buried in the internal folds along the midline of the occipital cortex in the back of the head. It is part of the visual system and seems particularly sensitive to social information, particularly tasks that require ToM.[6] This extra activity for the stag hunt extended to the neighboring region called the fusiform gyrus, a key part of the brain's face-processing system. Not only was the stag hunt associated with activity in the cuneus and fusiform gyrus, but there was a strong correlation between the increase in activity and the increase in payoff-dominant choices in the stag hunt.

It is important to point out that even though the subject was playing against another person in the stag hunt, the latter wasn't actually visible. In fact, the choices had all been generated by previous participants. There was no deception because we told the subjects this. So, why was there increased activity in face-processing and ToM networks?

A parsimonious explanation is that the subjects were mentalizing about their phantom players. This type of mentalizing requires two steps: imagining the presence of another person and then imagining what they would do. We already know that visual imagination repurposes visual perception circuits. When you close your eyes and imagine a sunset, you are activating your visual cortex even without input from your eyes. That was precisely what happened when people imagined someone playing the stag hunt. And if we take the magnitude of activation as a gauge of how hard they were working at this activity, it correlated with how often they went with the payoff-dominant choice. *This suggested that people who were better at mentalizing cooperated more often.*

Returning to Rousseau's parable, it is easy to see how evolution might have selected for mentalizing. When savage man realized his neighbors also had minds (who could in turn think about him), everything changed. Those who were able to harness this power to work together prospered, while those who didn't were left to hunt for rabbits alone. Models of evolution suggest that in any population there will be a mix of traits across individuals. If everyone hunted hares, the few people who worked together for stags would have a substantial advantage. But if everyone was hunting deer, the rugged individualist could still freeload off the group effort. A population will teeter between the two extremes until it finds what is called an evolutionary stable strategy.

Modern society favors cooperation. So much so we've created political, economic, and religious institutions to keep individualistic tendencies in check. Such institutions and their structures of rules are the only proven way to avoid the tragedy of the commons and protect the greater good. We'll see in subsequent chapters how these often become enshrined as sacred values.

As Rousseau realized, the benefits of living in society come at the cost of individuality. Mentalizing and ToM are powerful cognitive functions, so woven into the fabric of the human brain that no thought is truly your own. Whenever you're in a social situation, you have to constantly shift back and forth between your own thoughts and what you think another person is thinking. The mentalizing can even go one step further, to the point of simulating what they think about you—thinking about somebody thinking about you. In any social setting, then, you might have multiple versions in your head of who you think you are: your version and the version you think other people hold of you. And this is all because of humans' ability to think about other people thinking.

Which brings us, then, back to our original question: *Who are you?* As we've seen, with perceptual processes there is no right answer. The notion of a singular self is itself a narrative construction. There are the past-selves, the future-selves, the versions other people think of you, and all the flavors of you that manifest in society. In the next two chapters, we'll see how far this multiverse of yous goes and take a journey to see whether there is indeed an immutable core you.

CHAPTER 10

The Evolution of Groupthink

A S I SUGGESTED IN THE LAST CHAPTER, THERE IS AN evolutionary advantage to simulating others' minds, which comes at the cost of subjugating individuality. We are not born with this ability to mentalize or construct theories of mind, however. These skills come online in early childhood. At around the age of four, children figure out that other people can't read their thoughts. Whereas as adults we take for granted this division between our inner and outer worlds, as children the discovery of an inner life is revolutionary. Indeed, the recognition that one's thoughts are one's own is a crucial milestone on the way to full-fledged consciousness.

Following this discovery, a set of cognitive processes begins to unfurl. First, the child learns that they must speak to communicate their thoughts, and next realizes that they don't have to say everything that comes to mind. Some thoughts can remain private. From there, it's a short step to the realization that utterances don't have to precisely reflect thoughts. When you realize that your thoughts are private, you become powerful. You see that it is within your control to decide what to share with others. In fact, one's thoughts may be the last frontier of privacy; technology cannot (yet) eavesdrop on them, feed them to some algorithm, and sell you more of what you secretly crave. We cling to the notion that our thoughts are our own.

Unfortunately, this notion is a fiction, too. For where did those thoughts come from? Surely they weren't created entirely in the confines of our heads. Quite the opposite. In this chapter, we'll pull back the curtains on the Wizard of Oz in our brains and discover how shockingly commonplace our thoughts are. Not only that, we will see how the brain incorporates other people's opinions so readily that we mistake them for our own.

SOCIAL PSYCHOLOGY, THE STUDY OF HOW PEOPLE AFFECT each other's thinking, is a relatively young field. Just after World War II, the Nuremberg trials fresh in people's memory, there was a collective soul-searching to answer the question of why so many Germans went along with the Nazi ideology of genocide. Were they simply following orders, afraid to speak out? Or did the collective psychology of war somehow change how people saw the world? Such questions get at the heart of the ownership of personal narrative.

Solomon Asch, a psychologist at Swarthmore College in the 1950s, set out to answer this question in the decade following the war. Asch designed a series of ingenious experiments that showed just how powerful other people's opinions are in shaping our own perceptions. In Asch's most famous experiment, volunteers entered a lecture room in groups of eight for a test of visual acuity. They were all young men who prided themselves on being independent thinkers. The task Asch presented seemed trivial. As he explained, the men would be shown a series of cards with four lines drawn on them. The leftmost line was the template. All they had to do was pick which of the other three lines was the same length. Asch would then show an example to prove that the task was as straightforward as it sounded. To save time, he would go around the room and call upon each person to give his answer.[1]

In fact, there was only one subject in each round of Asch's experiment. Seven of the eight "volunteers" worked for Asch, who had instructed them to give the wrong answer for twelve of the eighteen cards. Asch was interested in what the one true subject did. Would he go along with the group, even when they were clearly wrong? Or would he go against the group and give the correct answer?

On their own, without a group giving wrong answers, 95 percent of the subjects performed flawlessly. But in the presence of the group, only a quarter of the subjects were able to maintain a perfect score. Everyone else gave in to the group pressure about a third of the time.

After the experiment Asch found that most subjects were aware of what they were doing, although they usually underestimated how often they went along with the group. Some accepted the group's judgment as an indication that their own perceptions were wrong. Others gave no evidence that they were even aware of their wrongness. In that sense, Asch's experiment didn't answer the question of why people go along with the group, but it made clear the relative ease with which supposed future leaders could be coaxed to act like sheep.

Asch believed that his subjects knew what they were doing. After all, the task was designed to be perceptually unambiguous. But with enough social pressure, most gave in to the fear of standing alone. And that was just in a lab setting, with nothing at stake except the opinions of their peers. It's not hard to imagine what people might do if the cost of sticking their neck out was even higher. Clearly, the experiment caused some mental gymnastics in terms of maintaining a coherent personal narrative. People had to concoct *some* explanation for their actions.

Asch's debriefing process hints at people's motivations for going along. Many of the individual subjects questioned what their own eyes were telling them. In addition, some claimed they altered their perception to conform to what everyone else said. If true, that would be a case of running the Bayesian equation in reverse to make perception conform to prior beliefs.

BY THE EARLY 2000S, SOCIAL PSYCHOLOGISTS HAD MOSTLY lost interest in the Asch experiments. These studies had been successfully replicated, and most researchers accepted Asch's explanation that most people did not alter their perception. But even Asch described the occasional subject who reported being wholly unaware of what they were doing. Maybe they didn't want to admit their sheep-like behavior, but there are insights to be found in such idiosyncrasies.

It was for this reason that I resurrected the Asch experiment in 2004. Instead of simple behavioral measures, though, we used fMRI to measure what was happening to perception itself.[2] As we saw in Chapter 4, prior experience shapes perception. We have also seen how stories serve as templates for personal narrative. Because stories originate from other people, I wondered, is it possible that other people's opinion fundamentally changes how a person sees the world? If so, then fMRI should detect changes in perceptual regions of the brain. If, on the other hand, conformist behavior occurs at the level of individual decision-making, we should see changes in decision-making regions.

Following Asch's script, we hired actors to play the roles of experimental subjects. When the real subject arrived at the MRI center, they saw that there were four other people, too. To further disguise the charade, the actors arrived at different times—some before, some after the real subject. We then gave instructions to everyone together. The task was almost as simple as Asch's. Instead of line segments, we used pairs of Tetris-like shapes on a computer screen. The participants had to decide whether the shapes were the same or different. On half the trials, each person would make a judgment on their own. On the other half, they would see what everyone else said. The real subject then went into the scanner to do the task, but unknown to them, the actors had been instructed to give wrong answers half the time.

Although the task was a little harder than Asch's, our participants did well enough on their own, getting 86 percent of the trials correct. When the actors gave the wrong answer, however, performance dropped to 59 percent, which was not statistically better than chance. When we debriefed the subjects, they offered a range of explanations for what they did. Some were not influenced by the group, while others went along almost 100 percent of the time. Most lay somewhere between these two extremes. They had a hazy remembrance of going with the group sometimes, but not at others.

When the participants did the task without the group's input, we observed widespread activity in the parietal lobes. This was expected because these regions receive inputs from the visual cortex and are central to constructing mental images. The parietal cortex is particularly active on tasks of mental rotation, and the exercise required the person to mentally construct images and compare them to the images their eyes were transmitting.

When we showed the subject the group's responses, two things happened. First, this external information resulted in activity decreasing in large portions of the mental rotation network in the parietal lobes. The decrease in activity suggested that when other people's answers were visible, the parietal lobes had off-loaded some of the mental work. This would make sense if the subject went along with what everyone else said and didn't do the mental rotation task for themselves. Second, when we divided the trials based on whether the person went along or not, we found that activity in the amygdala actually increased when the subject went against the group. The amygdala, separate from the cortex, is often activated during emotionally arousing situations. Clearly, standing up for one's beliefs was stressful. In most cases the subjects had no recollection of this, and may not have been consciously aware of the rise in stress at the moment of nonconformity. I suspected that in these circumstances the brain might try to shift its perception to what the group said, just to shut up that pesky amygdala.

OUR CONFORMITY EXPERIMENT WAS NOT PERFECT. IN THE time-honored tradition of experimental psychologists, we had deceived the subjects. And we hadn't paid them for correct answers, so there was no true incentive to report what they actually thought. The solution was to design experiments in which subjects were incentivized to give their best, most honest judgments on the task. The simplest way to do that was to pay people based on their performance. In 2008, I teamed up with Charles Noussair and Monica Capra, experimental economists at Emory. Both were well versed in designing experiments that didn't use deception while incentivizing subjects to reveal their true preferences. Instead of an Asch-type group conformity experiment, we decided to focus on another type of social influence: expert opinion.

Asch pioneered the concept of groupthink, but then his student Stanley Milgram gained greater fame (or infamy) for his electric shock experiments having to do with yielding to authority. In graphic fashion, Milgram demonstrated people's willingness to administer potentially fatal electric shocks to others when directed to do so by an authority figure.[3] Although the recipients' cries of pain were faked by actors, the field of social psychology turned its back on Milgram's approach because of its extreme use of deception and the mental anguish the participants experienced.[4] Charles, Monica, and I had no intention of resurrecting Milgram's experiments. The core idea, though, was similar: to explore whether an authority figure, either because they wear a white coat or carry an esteemed title, would have an outsized influence on perception.

We decided to focus on the bread and butter of economic paradigms: financial decisions. Specifically, we were interested in risky financial decisions. In economic jargon, a risky decision is any choice for which the outcome is not certain. A lottery is a risky decision. You can flip the ticket over and see the odds of winning. Then, on the basis of the size of the jackpot and the cost of the ticket, a person makes a choice of whether the price is worth the risk. The mathematical solution is simple: calculate the expected value (EV) by dividing the jackpot by the odds of winning. Assuming the ticket costs one dollar, if the EV is greater than one dollar, it could make sense to buy a ticket. Most people, though, do not treat money this way. Probabilities also figure into their decisions. And, if people don't treat monetary value objectively, then they are even worse with probabilities, tending to overestimate the odds of winning long shots while underestimating relatively sure things.[5]

Economists call this tendency *risk aversion*, and we set up our fMRI experiment to measure it in the subjects. While in the scanner, participants faced a series of lotteries. To ensure truthful responding, we told them that after the experiment, one of these trials would be chosen at random to be the one that "counts." If they had chosen the sure thing for that trial, we would give them the money. If they had chosen the lottery, they would have to play it using a random number generator.[6] Because the person didn't know which trial would be chosen, they needed to play every trial as if it would count. And because real money was on the line, they were incentivized to make choices in their own financial interest.

As we had done in the earlier conformity experiment, we had two conditions. In the base condition, the participant had to make decisions on their own. In the experimental condition, though, they were shown the recommended choice of an expert. As a distinguished professor of economics, Charles served as the expert. His advice followed a conservative strategy that was designed to maximize the chance of winning something. When the sure thing was above a certain threshold, Charles advised accepting it, regardless of the lottery amounts.[7]

The effect of the expert's opinion was subtle but clear. When we compared the decisions made with and without Charles's advice, we observed a shift in how people treated probabilities. They moved toward the expert strategy, becoming more conservative. The imaging data showed why.

These types of economic decisions are difficult. You have to do mental arithmetic and anticipate how you will feel about the possible outcomes. This level of cognitive simulation relies heavily on parietal and frontal lobes. However, when subjects went along with the expert's advice, we measured a decrease in activity in these networks. As in the Asch experiments, it appeared that people tended to offload their own mental work by allowing someone else to do it for them. When they went against the expert, the insula, which like the amygdala is associated with arousal, became more active, reflecting the discomfort of going their own way.

· Here we see indirect evidence of the porous boundary between self and other. Although each person made their choices of their own free will, they sometimes chose to offload that decision to the expert. Alternatively, you could say that the expert was in their head. Either way, the boundary between inner and outer worlds seems rather fuzzy. With ideas coming and going, it is easy to confuse which originate inside versus those that we absorb from elsewhere.

LOOKING AT THESE RESULTS, YOU MIGHT ASK WHY GOING against the crowd or the expert is so difficult. I've already mentioned the law of large numbers and how a group is more likely to render an accurate judgment than any individual.[8] Humans have a long history of evolving to live in groups. There is tremendous value for us, indeed, for any social species, in belonging to a group. The group provides protection, access to resources, and, as just noted, counterfactual opinions. When in doubt, ask a friend. Even better, ask several. The collective response will approach the correct answer. This is reflected in the well-documented accuracy of the average guesses of everything from jelly beans in a jar to the weight of cattle.[9] It may simply be that our brains are highly evolved to accept group judgments. You can either go with the flow or resist it, but resistance will send alerts through the brain's arousal system.

Going along with the crowd feels *good*, and this, too, would likely have been favored by evolution. If the crowd is generally right, then it helps to have a built-in reinforcement system for following it. A later fMRI experiment on financial advice found evidence of enhanced activity in the reward system for following advice.[10] Group opinion also serves as a form of information compression. If a person can offload decision-making to other people, then that saves on storage and processing in the individual's brain.

These are broad statements, of course, based on the average response of whoever happened to volunteer for our experiments. More interesting was the range of responses we received—what psychologists call individual differences. Average effects in experiments are great for describing broad tendencies in cognitive function, but deeper truths are to be found in answering the question of why some people are more, or less, prone to this phenomenon. Because everyone's personal narrative is based, in part, on what other people think, the question of individual susceptibility to conformity has a direct bearing on who you think you are.

To approach the question of individual differences, it helps to break down the different reasons why a person—either consciously or unconsciously—might alter their perception to conform to a group's opinion. In the years following Asch's original experiments, social psychologists identified two broad motivations for conformity. The first is the drive to obtain information. This follows from the law of large numbers and posits that some people attempt to improve the accuracy of the information they have by seeking the opinions of others. The second reason stems from the desire to behave in a way that is socially acceptable, which is called a normative influence.[11] Presumably, these two forces would manifest in different brain systems and their relative potency might explain an individual's susceptibility to conformity.

When it comes to a process like conformity, brain imaging holds an advantage over traditional behavioral testing methods. In any given circumstance, both informational and normative influences might affect a person's decisions. But most people (especially in the individualistic West) pride themselves on independence. Few are willing to admit that their decisions might be a result of following the herd. With brain imaging, the answers to which influences might affect a person's decisions don't depend on a participant answering that question honestly. We can simply observe which brain systems come online when they make a decision.[12]

With this caveat, we set out to discover whether conformity operated on brains primarily through perceptual changes, as was suggested in our previous Asch experiment, through social forces, or through both. We hoped that parsing out the relative contributions of these different brain systems would also explain the individual differences. Figuring that older adults might be too jaded to participate in this type of experiment, we decided to focus our attention on adolescents. Who cares more about what their peers think than teens? So went our thinking. We set out to recruit teenagers between the ages of twelve and eighteen.

Setting a task was a bit more of a challenge. The mental rotation task might favor those with video game experience or be too difficult for some of the younger kids. Plus, because there was always a right and wrong answer, it was primarily a test of ignoring one's own judgments. We needed something that didn't have a right or wrong answer, but for which people still had strong opinions. We quickly homed in on music. Everyone has songs they like and songs they dislike. Teens, in particular, have always placed great importance on the music in their lives. The question was whether they liked certain songs because they were popular, that is, because other people said they were good, or the kids came to their musical preferences on their own.

To set up an experiment around this question, we needed music they hadn't yet heard.[13] So, we turned to the internet. In 2006, when we began the experiment, the dominant online music site was MySpace.com. Originally conceived of as a social networking site in 2003, MySpace was created as an alternative to Friendster at a time when Mark Zuckerberg was just starting to code the early version of Facebook from his Harvard dorm room. By 2006, MySpace was much more than a social network. Because of the ease with which the site let musicians put out their work, MySpace had become the primary outlet for new artists. Conventional record labels took notice and began tracking downloads to seek potential new artists to sign. Although we couldn't be sure that our participants were also using MySpace, the site was our best source for undiscovered music.

Sara Moore, the lead research specialist in our lab, volunteered to scour MySpace for content. Anticipating that the participants in the imaging study would have a range of musical tastes, Sara made sure to find songs from six genres: rock, country, alternative/emo/indie, hip-hop, jazz/blues, and metal. The only requirement was that the artist not be signed to a label. Sara did all her trawling during a two-week period, creating a snapshot in time of what people were listening to. She narrowed her findings down to 120 songs, 6 from each genre. As a measure of popularity, she noted how many times each song had been played, which ranged from 876 to 1,998,147, and converted this to a five-star popularity scale. To put this in perspective, plays by the top signed artists approached 100 million.

There was no way we could play 120 songs in the scanner. That would have taken hours. But most popular songs are highly repetitive, and you can get the gist from a fairly short clip. We boiled each song down to a fifteen-second snippet that included either the hook or the chorus. At the beginning of the fMRI session, the participant ranked the genres from one ("the type you like best") to six ("the type you like the least"). Because we were trying to see how a song's popularity influenced an individual's response to it, we assumed that any such effects would be small and unlikely to make a person like something in a genre they already hated. So, we used songs from only their top three genres.

In the scanner, the participant listened to each clip twice. After the first time, they gave a rating of how much they liked the song on a five-star scale. We then played the clip again, but this time, we showed them how popular the song was. After that, the participant had the opportunity to revise their own rating. As a control condition, we held back the popularity display on one-third of the trials. For these, the subject provided a second rating without the influence of other people's opinions.

First, we looked to see whether a song's popularity affected the kids' own ratings. On the trials where we hid the song's popularity, subjects changed their rating 40 percent of the time. But when we revealed the popularity, this increased to 80 percent. To quantify the susceptibility of an individual, we calculated a conformity index. The idea was that if the initial rating was lower than the popularity score, then a conformist would raise their rating. Similarly, if the popularity score was lower, then they would lower theirs, too. As expected, there was a wide range of conformity scores. To ferret out the source of this individual difference, we looked at two factors: gender and age. Gender wasn't a significant factor, but age was, with younger subjects having a higher conformity index. This suggested that young teens were not as set in their musical preferences, or they were more susceptible to peer pressure effects, or both. Either way, the age effect pointed to the malleability of preference during a critical period of development.

We still wanted to know whether this meant the kids actually liked a song more or less on the basis of popular opinion or they maintained their preference internally and changed their rating to go along with the crowd. As we did in the Asch experiment, we first examined the fMRI data for brain regions activated by listening to the songs. As expected, a widespread network including areas involved in auditory processing and attention was activated. And unlike in the Asch experiment, which involved geometric shapes, pleasure and aversion played a key part in our musical study. We confirmed that there was a correlation between the first likability rating and the brain's response in the individual's reward system. Activity in the heart of the reward system—the caudate nucleus—increased when a person rated a song highly and decreased when they didn't. We found no evidence that discrepancy between an individual's rating and a song's popularity changed this reward response. This meant that, on average, the popularity rating did not change how much a person liked the song.

When we focused on only those songs for which an individual changed their rating, we found decreased activity in the caudate. The greater the discrepancy between the person's initial rating and the popularity, the greater the decrease in the reward circuit. Correlations, though, can be considered two ways. The other way to interpret this result was as a boost in caudate activity when a person saw that they had rated a song the same as popular opinion. We saw the largest reward response when a person rated a song highly and then saw that everyone else did, too. Such a finding would be consistent with a person obtaining a social reward from conforming to the group or, conversely, avoiding the pain of being a contrarian.

Satisfied with the results, we published this teenage music experiment, and I thought that was the end of it. I moved on to other projects. Meanwhile, evidence continued to accumulate for the merging of social rewards and personal preferences within the brain's reward system. Using a similar experimental design, a group of researchers in London also found a boosting effect in the reward system for songs that were highly rated by other people. Additionally, they found evidence for activation of the pain system in the insula when a subject's ratings diverged from others'.[14] Such effects were not limited to music. The boosting effect of social conformity was similarly observed when participants rated the attractiveness of faces and then saw what other people thought.[15] In fact, social popularity seemed to affect the reward and pain systems of the brain across many types of decisions—everything from simple purchases to the price of wine and stocks.[16]

Three years later, I was watching *American Idol* with my two daughters. The show was in its eighth season, but I hadn't paid it much attention. My daughters, nine and ten years old at the time, loved the show. Kris Allen and Adam Lambert had been battling it out all season. It was down to the semifinals. I was only half listening, mainly enjoying the girls' enthusiasm for the show. My ears perked up, though, when Kris Allen began singing a song that seemed familiar. It was "Apologize," by OneRepublic. I was completely unaware of how big the song and the band had become. After Allen was done, I told my daughters that I knew that song. Their looks said, *Duh*. We had used "Apologize" over three years before Allen belted it out on *American Idol*.

This experience prompted me to wonder whether there was anything in those teenagers' brains, scanned years before, that could have predicted the success of this song. And what about all the other artists we had sampled? How many other OneRepublics had hit the big time?

FOCUS GROUPS HAVE LONG BEEN THE FOUNDATION OF marketing campaigns. You take a group of people who are representative of a demographic and solicit their feedback on a product. Feedback can take the form of unstructured discussions, questionnaires, or A/B product comparisons, where a person has to pick which option they prefer. I had stumbled into something entirely different, though. When I set up the music experiment, my intention was to study the neural basis of conformity. But I had also created, without intending to, a neurofocus group. If we had obtained a representative sample of teenagers back in 2006, it was theoretically possible that their brain responses to music would have been representative of their peers in the larger music-consuming world. And if that was true, the brain responses of our neurofocus group might correlate with some metric of music sales.

Three years had passed between when we had collected brain responses and "Apologize" hit *American Idol.* That should have been plenty of time to collate sales data. Many of the bands had either disappeared or, at least, there was nothing in a sales database for the particular song we had used. That seemed about right. There was probably good reason that these artists had not been signed to a label at the time. Many of the songs were terrible.

But 87 of the 120 songs had sales data, and that was enough to determine whether sales correlated with anything in the teens' brains. This was exciting because even though we were attempting to look for correlations of sales with brain responses, *the sales hadn't yet occurred at the time of brain imaging.* If we went back in time to the day of scanning, we would have been looking into the future, asking whether we could predict future sales on the basis of brain responses.

Before examining the brain data, we checked to see whether the participants themselves could have predicted a song's success through their likability ratings. We found no such correlation. Simple reports of likability, in other words, were not good predictors of commercial success. But when we looked for correlation in the brain, we found what we were looking for. Sure enough, the activity in the nucleus accumbens—the heart of the reward system—did correlate with sales.[17] Further analysis showed that this correlation was driven by the coordination of activity between the nucleus accumbens and the part of the frontal cortex that is folded on the underside of the brain, just over the eyeballs.

It was exciting to see the data, but just because we found a correlation doesn't mean we could have predicted a hit. By industry standards, a hit is a gold record, and that takes 500,000 album sales (1 million for singles). Only three songs from our experiment met that threshold, including "Apologize." Picking those three songs out of the original 120 was not possible, but if we relaxed the definition of what constituted a hit, then patterns began to emerge. What if we considered a hit to be a song with 100,000 sales? What if we dropped it to 10,000? The sweet spot for our dataset was around 30,000. At that threshold, the kids' brains could be used to predict with 95 percent accuracy which songs would go on to sell more than that, while still correctly predicting 80 percent of the songs that wouldn't.

You might wonder whether brain imaging was necessary to do this. At the time, Google and others hadn't yet reached the level of data mining they currently employ. Early research had shown that the frequency of internet searches could be used to predict revenue for movies and video games, but less so for music.[18] Maybe there was something idiosyncratic about music that made it hard to predict hits. More likely, it's the sheer volume of songs that are available, which is orders of magnitude more than the number of movies or video games released each year. If so, that would make our results even more impressive.

I still find it remarkable that we could scan the brains of a miniscule sample of the music-consuming population and predict something about how popular songs would become. The findings suggest that our brains are more similar to each other than we imagine. We like to think that we are individuals. In Western societies, individuality is central to our personal (hero) narratives. We believe that our thoughts are our own. That is clearly not the case. There is enough in common between thirty-two teenagers' brains and the rest of the population that we could predict some degree of commercial success of music. At a minimum, it means that we share reactions to popular media that are uncannily similar.

You could argue that this finding is no surprise. Our musical preferences and financial decisions are, after all, subject to societal expectations and cultural conditioning. But maybe these shared responses run deeper than that. In the next chapter we'll look at whether these social conformity effects extend to more weighty matters such as religion, political affiliation, and even your core identity.

CHAPTER 11

Moral Backbone

W E NOW HAVE AMPLE EVIDENCE THAT THE BOUNDARY
between one's own thoughts and what other people think
is more porous than we might realize. And because we live together in society, we constantly have to navigate the competing
tugs of self-interest and common good. Even when these interests
are aligned, as in the stag hunt, it can be difficult to work together.
The ability to put oneself in another's shoes helps to some extent.
Indeed, people who are better at mentalizing seem to cooperate
more—at least in stag hunt scenarios.

Yet society shapes the human brain beyond mentalizing. Because living as part of a community versus being an outcast has
historically been a matter of life and death, evolution has strongly
favored people who can work together. Some of this is biologi-
cally hardwired, in other words. But society holds itself together
through cultural transmission of codes of conduct, inculcating its
members with a set of morals and tempering selfish impulses for
the common good. I have already touched on the importance of

first stories in the development of a child's moral structure, which itself forms the backbone of their personal narrative. In this chapter we'll delve further into how the power of society continues to shape who we think we are.

There is a fundamental conflict between the individual and society. The developmental psychologist Erik Erikson framed it in his fifth stage of psychosocial development as the tension between identity and role confusion.[1] This is the period of life when an individual questions who they are and what their role in life is. Although this tug-of-war tends to peak in adolescence, it is a conflict that reemerges throughout one's life. As circumstances change and people and jobs come and go, it is a natural part of the human experience to return to the question of who you are. More precisely, who you *think* you are.

Morals and sacred values form a major part of our conception of self. Because they're laid down so early in life, they often form the foundation on which the other components of identity are built. These other pieces—including your occupation and your personal relationships—are important, too, but they accrue over years of living. When life is going well, most of us find little need to take a personal inventory. It's when the world throws a wrench into the works that we're prompted to look inward, if only to recalibrate who we actually are. Do you think of yourself in terms of your job? Your accomplishments? Your partner? Your children? Or do you think of yourself in terms of core values? It is sobering to realize, as many do in a crisis, that everything except your values can be taken away.

Indeed, many people will tell you that morals and sacred values define the core of their identity. Take the Golden Rule, often recounted as "do unto others as you would have them do unto you." If I had to offer my children one rule to live by, this would be my choice. It is an excellent strategy in almost all facets of life. It requires some degree of mentalizing, and so through repeated use builds the necessary human skill of empathy and consideration for others. But when parents teach the Golden Rule to their children, it is often presented as a commandment. The earliest form of the Golden Rule, dating to ancient Egypt in around 2000 BCE, was worded as such. It appears so frequently in historical texts, one wonders why our ancestors had to beat us over the head with it. The reason is because, despite its utility, it is an exceptionally hard rule to practice. The social psychologist Jonathan Haidt suggests: "Moral systems are interlocking sets of values, virtues, norms, . . . and evolved psychological mechanisms that work together to regulate selfishness and make coordinated social life possible."[2] It sounds like a commandment because it is a commandment. Without the Golden Rule, society would descend into chaos.

The Golden Rule exemplifies the inherent paradox of holding sacred values. Though we may identify with them as core constructs of our identities, it is also clear that they didn't come out of nowhere. Sacred values are implanted in our brains by parents and society as a whole, becoming a piece of the self delusion. The ubiquity of many sacred values illustrates how our notion of self is porous, accepting what authority figures tell us is for the good of society. But as we saw in the last chapter, individuals are likely to vary in their proclivity to absorb sacred values. Some people are rigidly moral, perhaps to the point of insufferability, while others may appear to be amoral and completely self-interested. Where you fall on this spectrum is an important factor for determining how malleable your personal narrative might be.

A SACRED VALUE APPEARS SELF-CONTAINED, BUT IT ALSO acts as a highly compressed representation of a complex narrative. These narratives come from the stories we absorb growing up. When it comes to sacred values, I think of the Ten Commandments—proscriptive and carved in stone. Although the Ten

Commandments stand alone as a monolith of the Old Testament, whether a person views them as sacred depends on how the individual engages with them and the culture they live in. The big sacred values tend to cluster around harm to others (Thou shalt not kill, Thou shalt not commit adultery). Depending on one's beliefs, less weighty sacred values can filter down to seemingly mundane daily choices like eating kosher or halal foods, patronizing Christian businesses, or investing in socially responsible mutual funds.

Although sacred values are important to who we think we are, they are also devilishly hard to study. You can't simply ask a person what is sacred to them and expect an accurate response. That's because people don't like to talk about them, at least honestly. Many people espouse core values, but then go on to engage in behavior that goes against everything they supposedly believe. The Ten Commandments exist because the things they proscribe are exactly the activities people have the selfish impulse to engage in, and doing so would tend to destabilize society.

If the study of sacred values resists the direct approach, then maybe an indirect route can yield insights into what people consider the core of their identity. I believe that brain imaging can provide such a window into people's psyches. Virtue theory, a subfield of ethics, suggests that there might be two very different ways in which sacred values are represented in the brain.[3] The utilitarian philosophers John Stuart Mill and Jeremy Bentham thought that moral decisions should be guided by the principle of the greatest good for the greatest number.[4] If an individual considers a sacred value as a cost-benefit analysis, as in, *If I do/ don't do this, I will go to Hell*, then this should manifest as activity in brain structures associated with reward and punishment. The alternative, the deontological view espoused by Immanuel Kant, holds that a sacred value is nonnegotiable.[5] It is a rule. If true, then the associated brain activity should appear in regions linked to rule processing. How a person represents their personal sacred values—whether as a cost-benefit calculation or a rule, such as always stopping at a stop sign—is crucial to understanding who they are. These differences don't make a person more or less moral, but they dictate how a person processes certain types of information. The importance of these differences will become even more apparent in the next section, when we delve into rewriting our narratives.

ALTHOUGH I HADN'T ORIGINALLY THOUGHT OF MY LAB'S research as relevant to sacred values, in fact, I had been studying the utilitarian side of the coin through our work on economic decision-making. But I still knew nothing about how the brain might represent the deontological side of the coin. For that, I turned to Scott Atran, one of the world's leading experts in how violation of sacred values is at the root of terrorist acts. I met Atran in 2007 at a meeting of researchers funded by various Department of Defense agencies. The topic of his talk was how he reconstructed the network of terrorists responsible for the 2004 Madrid train bombings. How did he do it? He interviewed the friends, wives, and families of those responsible.

Atran concluded that the bombers were rather ordinary people. There was nothing pathological or deranged about them. They were like many young men across Morocco and Tunisia, looking for economic advancement but not finding much opportunity. That they fell in with each other was happenstance. They shared a love of soccer and a hatred of the West for blocking their futures. And although their original grievances might have been economically based, they transformed into sacred ones through the linkage to membership in their group. The terrorist mindset resulted from small-group dynamics coalescing around a common theme, such as humiliation. But what was responsible for that transformation from a mere grievance to an idea that someone was willing to kill and die for? Before we could answer that question, we needed to know which side of the sacred coin—utilitarian or deontological—a person was using. And this is where we thought brain imaging might help.

Atran had an easygoing, nonjudgmental manner that made him a pleasure to talk to. People just spilled their secrets to him like he was their bartender. It was a skill that had saved his life, for he had interviewed people in the most dangerous places in the world, places like Afghanistan, Syria, eastern Turkey, and Indonesia. But reproducing this sort of interaction in the MRI scanner was not in the cards. The MRI is a sterile, clinical environment. Many people say it feels like a coffin. Besides, it isn't possible to obtain useful brain data while people are talking, because speech causes too much head movement. We couldn't use behavior either—unlike most types of simple decisions that we can study in the scanner, like monetary choices, those driven by sacred values don't typically have an easily observable behavioral manifestation.

We needed to probe people's sacred values in a way that didn't require a physical demonstration. If someone claims that Jesus was the son of God, and that is an absolutely sacred belief to them, what could they possibly do to prove it? Nothing. Still, Atran and I felt that there must be a brain signature that could distinguish between a true believer and a pretender. One of the questions that Atran used in his interviews attempted to reveal under what conditions people would compromise their values. For example, if the Sabbath was sacred, but if your mother was deathly ill, would you take a taxi to the hospital? Most people would make an exception, recognizing that familial obligations— especially emergencies—are sacred, too. This example shows that it is difficult but acceptable to make choices between sacred values. It is never the case, though, that a sacred value could be exchanged for something nonsacred—like watching the Yankees play on Saturday, no matter how big a fan you are. Atran's technique probed hypothetical trade-offs like these. The answers gave clues to what a person believed was sacred without requiring them to demonstrate it.

This kind of dialogue would not be possible in the scanner, but we came up with an alternative technique that prompted similar responses in a more streamlined manner. Our solution was to present people with statements designed to elicit their position on various sacred values. Imagine presenting a Christian who believes that Jesus was the son of God with the statement: *Jesus was the son of God.* If you truly believe that, such a claim should evoke some sort of self-affirmation, maybe even a good feeling. Contrast that with the opposing statement: *Jesus was not the son of God.* You might find that offensive, and that reaction would be associated with neural responses associated with disgust. An agnostic person, on the other hand, might not display much of a brain response to either statement. If the value isn't sacred, then there is nothing to affirm or be offended by.

This approach was at the heart of our plan. We would simply present subjects with statements phrased in the second person. Some would be designed to push and pull on sacred values—at least, those we presumed would be sacred—while others would be decidedly nonsacred. Nonsacred values are properly termed *preferences.* Examples include whether you like cats or dogs, coffee or tea, Coke or Pepsi, PCs or Macs. Mere preferences, we hypothesized, should not result in much effect on brain activity. If anything, preference statements might elicit activity in the reward system as people weigh the costs and benefits.

The notion of costs and benefits, though central to utilitarian decision-making, is anathema to deontological (belief-based) decisions. We hoped that we could tell the difference between these two types of decisions by monitoring the reward system and the rule-processing systems of the brain as people considered the state-ments we presented to them.

I was quite fond of our plan, but Monica Capra, a professor in the Economics Department and longtime collaborator (with whom I also worked on the expert-advice experiment from Chapter 10), didn't buy any of it. She pointed out the lack of a gold standard. Could we really believe that a person's brain activity alone was suf-ficient to tell whether something was sacred to them? I took Mon-ica's complaint to heart. We needed an independent measure of sacredness.

Monica had taught me earlier about a concept economists called "incentive compatibility." Basically, if you're going to probe people's preferences, you had better make sure that they are incentivized to give true responses; otherwise, people could make anything up, at no cost to themselves. For simple prefer-ences, this often meant using some sort of auction mechanism. A person's preferences are revealed by how much they are will-ing to pay for something. There are many types of auctions, but one that had gained favor among experimental economists was called the Becker-DeGroot-Marschak (BDM) auction.[6] With this setup, a participant submits a bid for an item, ranging, say, from $1 to $100. Then, a random number is chosen between 1 and 100, representing the asking price. If the bid is greater than the random number, the participant pays the amount of the random number and receives the item. The BDM auction is considered incentive-compatible because, unlike an eBay auction, there is no reason to bid more than you are willing to pay. Conversely, un-derbidding decreases the chance that you will get the item. The best strategy is to bid exactly what something is worth to you.

It's a good strategy for material objects, but this approach posed a challenge for sacred values. By definition, sacred values are priceless. Suppose, I suggested, that we ask participants a series of forced-choice questions: *You believe in God* versus *You don't believe* in God; *You are a Mac person* versus *You are a PC person*. Then, at the end of the experiment, they would sign a document attesting to their choices. It would amount to an inventory of personal values. Monica rightly pointed out that such a questionnaire still wouldn't obligate a person to having those values. Even if it didn't, I argued, there would be a personal cost to signing a statement that was contrary to your beliefs—who you thought you were. What I had in mind was a measure of integrity, which would act as a sort of proxy for sacredness. Theoretically, a person should have more integrity for a sacred value than for a mundane preference.

But there was a twist. In order to gauge the strength of integrity, we would hold a BDM auction for each of a person's values. After the participant made their personal value selections, but before signing the document, we would offer them money to change each of their answers. If they didn't want to sell a particular answer, they could opt out of the auction for that item. For the others, they would set an ask value, and we would roll two 10-sided dice to obtain a random bid value between $1 and $100. Submitting an asking price meant that a person was willing to exchange an item for money, and the item was, by our metric, not sacred. Opting out was taken as a sign of sacredness.

Monica was still skeptical. What would stop a person from auctioning off every item and taking the money, whatever they truly believed? So I proposed a solution: to come up with statements that no one would attest to, no matter what the price.

We pressed other lab members into the effort. I should have known better. During lab meetings, the usually boisterous group was uncharacteristically silent. Nobody wanted to say what was sacred to them, or worse, articulate words that were the antithesis of sacred: the profane. The lab members were a diverse group. Of the eight regulars, we were evenly split between men and women; two identified as queer; one as Black; one as Latina; multiple as Jewish. Still, there was no way to sugarcoat it. We needed to get

comfortable saying what was sacred to us and which statements
were offensive.

It took weeks, but eventually the group got comfortable enough
with each other to speak openly about what was taboo in public.
We quickly homed in on hot-button issues: sanctity of life, race,
sex, politics, and religion. Even though this was back in 2010,
not much has changed in this regard. Here is a sample of value
statements that we thought would have a high likelihood of being
considered sacred. Of course, we would also present the opposite
statements. Because those are the most offensive, I will leave those
to your imagination.[7]

- You do not like to hurt animals.
- You think that marital rape is a crime.
- It is not okay to use nuclear weapons on civilians.
- You would not cheat on your spouse even if there was no
  chance of getting caught.
- You are not willing to kill an innocent human being.
- You think it is not okay to sell a child.

Nonsacred statements of preferences were easy and included
the likes of

- You are a Pepsi drinker.
- You are a dog person.
- You have a favorite color M&M.
- You prefer to watch football over basketball.

Not only were these statements of personal values, they were
loaded with history, or more pejoratively, baggage. The sacred
ones particularly so. Each statement compressed a massive amount
of cultural, religious, and personal narrative into a single sentence.
Before we invested time and expense in the imaging experiment,

we tested the statements using an online survey with a group of 391 people. Atran's colleague Jeremy Ginges at the New School in New York set it up. We wanted to know the proportion of people selecting each statement in the pairs and whether, hypothetically, there was a price for which they would change their answer. The results confirmed that our statements covered a wide range from the utterly mundane to the profoundly sacred.

Confident that we had a decent slate of values, we proceeded to the imaging phase. We cast a wide net for volunteers. Students, of course, were easy to find, but they were not very representative of the general population. University students tended toward the secular and were less likely to hold sacred values. But this was Georgia, and we suspected finding churchgoers who held many values sacred wouldn't be a problem. We eventually recruited forty-three people for the fMRI study. Seventy-four percent of the group believed in God, but only 60 percent said that religion was an important part of their identity. Seventy-four percent identified as Democrats. Sixty-three percent supported gay marriage (this was before *Obergefell v. Hodges*, the 2015 Supreme Court ruling that legalized same-sex marriage).

Although the auction format wasn't a perfect measure of sacredness, it served as a reasonable proxy for integrity. For each participant, we grouped their results into two categories: statements with bids and those without. For grouping purposes, we considered the chosen statement and its opposite as sacred or as nonsacred. Then, we analyzed the brain responses obtained during the passive reading phase, comparing sacred to nonsacred statements. Keep in mind that the person did not know that they would be bidding on these items when they saw them for the first time, so they could not have hedged their reactions at that point.

The neuroimaging data revealed activity in a network of three regions: the left prefrontal cortex; an area toward the back of the brain at the border between the temporal, parietal, and occipital lobes called the temporoparietal junction (TPJ); and the right amygdala. The prefrontal region is next to areas normally associated with language and speech but also known to be active in rule processing.[8] Similarly, the TPJ has been previously implicated in moral judgments.[9] Together, the appearance of these two regions implied that the volunteers processed sacred statements as rules, not decisions. We confirmed this later by asking them whether they made their choices based on rights and wrongs or costs and benefits.

The amygdala activation, however, was present only for the opposites of the chosen sacred values. Of course, these would be the most offensive statements. The amygdala is an emotional hair trigger in the brain. It is frequently active in high-arousal situations. Violations of sacred values evoked strong emotional reactions ranging from disgust to humiliation. The amygdala activation confirmed what Atran had inferred from his interviews throughout the world. Still, it was reassuring that, at some level, we could duplicate these types of reactions in the laboratory setting. And because the brain responses were evoked simply by passive reading of the statements, they were not dependent on eliciting choice or even waiting to observe a person's actions. It was truly a window into a person's soul, or at least the price of it.

We wondered whether there were other clues to the strength of a person's values. We checked for correlations with personality. Nothing. Not even the strength of a person's religious beliefs was correlated with the brain response. Instead, we found the prefrontal response to sacred values was correlated with a participant's level of involvement in group activities. This included any organized volunteer activity, such as sports teams, political organizations, religion, or clubs. The more a person was involved with group organizations, the stronger their brain response to sacred values.

It is possible that people with strong personal values also tend to be more active in groups. Or active participation in a group might nurture the cohesive tendencies in a person's brain, that is, an individual's susceptibility to groupthink. Simply belonging to an organized group—whether it is a religion, a sports team, a political group, or a book club—requires an individual to consider other people's opinions. Groups require individuals to adjust their values and points of view to conform, in part, to the social norm.

What is surprising, and frightening, is that even membership in secular groups may nourish conformist processes in the brain to the point that they spill into the realm of the sacred, in effect, making people more dogmatic about their values than they might have been on their own. As Atran discovered, the Madrid bombers not only attended mosque together but also played soccer together. We must acknowledge the very real possibility that being an active member of society, which requires engagement with a range of groups, may also crystallize values to the point that they become nonnegotiable, in effect, sacred.

The polarization of the United States offers a prime example. Republicans and Democrats have coalesced around their own set of sacred values. Democrats hew to abortion rights, gun control, and universal health care. Republicans cling to pro-life, the Second Amendment, and low taxes. By design, these sacred values are mutually exclusive with the other side's values. It forces people to choose sides. Is it any wonder that many people choose not to choose?

BUILDING ON OUR STUDIES OF THE EFFECT OF POPULARITY on music preferences, we asked whether the same factors were at play in people's moral values. Let's say you believe in God. This is a strongly held belief and one you consider to be a core value of your identity. Now imagine you find yourself in the company of a group of atheists. Conversation turns to religion and everyone goes around expressing their conviction that there is no God.

Everyone looks at you, awaiting your response. What do you do? (For the atheists, simply reverse the roles and see how it feels.)

If it is truly a core value, then you are probably thinking that you would stand up for yourself and say what you believe. But are you sure? Maybe you would hedge slightly, saying you believe in God but you're not really sure. Or, for the atheists considering how they would feel, maybe you would hedge and say you're agnostic. This was exactly the situation we aimed to create in the experiment.

We used the same set of questions that we had employed in the sacred values experiment, but with a twist.[10] As before, subjects in the MRI scanner were presented with a series of statements. First, they watched them passively, and then engaged in a choice phase. Next came the twist. In the third phase, we displayed the subject's choice for each of the values along with a thermometer graphic indicating how many of the other participants shared that value. This was followed by a yes/no question of whether there was any amount of money they would accept to switch positions. The subjects then came out of the scanner and participated in the auction, as before.

In total, seventy-two people participated in this experiment, a large number by any standard for an imaging study. First, we calculated the impact of the popularity meter. We did this by looking for a correlation between the height of the meter and the bids participants submitted during the auction (while also controlling for their average bid). This let us estimate how sensitive each subject was to the popularity gauge, which we called a conformity score.

The imaging data again showed activation to sacred values in the same region of the left frontal lobe. But what was interesting was that the magnitude of activation was negatively correlated with the conformity score. People with low conformity scores, meaning their bids were unaffected by popular opinion, had the highest activity in this region of the frontal cortex. People with high conformity scores had the lowest activity. The results suggest that we can use the activity in the left frontal lobe as a marker for *deontological resolve*—the willingness to stand by one's values in the face of contrary opinion. Call it a moral backbone.

WHEN ATRAN AND I COMPLETED THE SACRED VALUES EX- periments, I felt confident we had elucidated a basic schism in the brain's decision-making mechanisms. Reward systems dealt with utilitarian calculations, and rule systems handled sacred values. I still think that is fundamentally true.

But as time has gone on, I've come to appreciate that sacred values are far more complex than mere rules. They encompass whole ideologies, be it political or religious. Sacred values are some of the most highly compressed narratives in our brains. Take the abortion divide. It is easy to say, "I am pro-choice," or "I am pro-life." The reality for most people, however, is a nuanced consideration of the circumstances under which they think abortion is okay and when it is not. Of course, that kind of thought requires time and care- ful consideration, including the weighing of consequences for the mother, the child, one's religion, and the local social norms. Iden- tifying with one side or the other simplifies matters significantly.

Sacred values are more than they appear to be. They represent narratives that have evolved over thousands of years. Moreover, they serve their purpose by tempering our sometimes selfish im- pulses so that we can live together in a collective society with rela- tive harmony. Finally, although it may seem counterintuitive given today's political divisions, they remind us of how similar we are.

The upshot of all this is that evolution has given us brains adapted to living with other people, and that means that very few thoughts are truly originally ours. There is no doubt that work- ing together is usually superior to going it alone. The challenge is always to temper our own individual interests against the public good. The ability to see things from other people's perspectives and to absorb their opinions as our own are two skills that evolution has given us to help us work together.

At this point, considering all these constant and hardwired external influences, you may be wondering what—if anything—constitutes a true you. Yet, as I hope you're realizing, that question in its limited focus on the individual misses the point. The only way to get to something like an answer is to expand the definition of the self. Each of us is much more than whatever is contained in the physicality of our bodies. I am not referring to something ineffable like a soul. What I'm suggesting is that a conception of our self makes no sense if we don't include all the interactions we have with other people and with the world itself. In this model, "you" places the physical you at the center of a network of connections, and those connections spread out like roots from a tree, touching everything around you.

A virtuous human being, then, is greater than their own egocentric narrative. Even if one hews to the hero's quest, the hero gives back to society by telling other people what they have learned on their journey. Such a person acknowledges the give-and-take between their own narrative and the norms of society. Of course, there will always be people who thumb their noses at what other people think. Sometimes this is productive, like when an iconoclast invents new technology, but other times, it is antisocial and sociopathic. In the next chapter, we'll take a close look at the brain of such a person—someone who has violated the most fundamental of sacred values.

CHAPTER 12

The Banality of a Brain

A S WE'VE LEARNED SO FAR, MUCH OF WHAT GOES ON IN our heads has been culturally programmed and crystallized in our brains as a personal moral compass. Strictly speaking, they are not our own thoughts because they are shared with everyone else.

Well, almost. In any society, there are outliers who exist outside the cultural and moral norms: loners, miscreants, sociopaths. When Scott Atran and I began brainstorming about sacred values, the possibility of scanning a terrorist's brain was legally and logistically impossible. In Atran's experience, most terrorists were psychologically normal people who held an extreme belief. According to Atran, a terrorist's brain would look no different from anyone else's. Long before I met Atran, I had, in fact, had a chance to test this theory. Any such differences found might hold clues for those who want to break free from the bounds that society imparts on its members (nonviolently, of course).

In the twenty-plus years I have been studying the human brain, I must have put over a thousand people in the MRI scanner. I remember the first time I crawled in, just to see what future volunteers would go through. I remember the first time we hooked up two scanners together so that we could scan two people at the same time while they played a game with each other. I remember training my favorite dog to go in the scanner, and the swell of pride and elation I had when we completed the first scans of her brain while she was completely awake and unrestrained. But apart from those highlights, I can scarcely remember the personal details of any of the thousand people I have scanned.

Except one. I will never forget.

Demographically, he was no different from the usual volunteers. Male. Age: twenty. Average height and weight. He spoke with an accent that I had come to recognize as typical of Georgia, more rural, and clearly inherited through several generations, probably pre–Civil War. Most of the college students who came from Georgia eventually softened their accents while on campus, the inevitable result of assimilation into a multicultural community filled with accents from all over the world. But this participant was not a student.

The orange jumpsuit was a giveaway. That, and the handcuffs and the leg restraints. Plus, most volunteers don't require two armed deputies as guards.

IN THE LAST CHAPTER, WE EXAMINED SACRED VALUES AND how they are represented in the brain as both rule-based values and highly compressed narratives. It's one thing to study such morals in the abstract, but it is another thing entirely to put your value system to a real-world test. Most people will never find themselves in a situation where someone else's life hangs in the balance, let alone where they could kill another human. Indeed, the sanctity of human life approaches a near-universal sacred value.

Moreover, murder is not very common. According to the CDC, in 2018 there were 19,141 deaths officially deemed homicide in the United States.[1] That comes out to 6 deaths per 100,000 people a year. For comparison, heart disease kills 200 people per 100,000 each year; accidents, 52 per 100,000; and suicide, 14 per 100,000. Of course, this varies by age and socioeconomic status, but the point is that not only is murder rare, it is unlikely that you will ever know a murderer. I certainly never thought I would.

In early 2002, I took a call from a criminal defense attorney who had heard of our recent work with fMRI. Daniel Summer had been assigned the case of a client who stood accused of murder, among several other charges. Because of the circumstances of the crimes, and this being Georgia, Summer's client would face the possibility of execution if convicted. Summer was convinced that something was off about his client, that he didn't seem to process information normally. I suggested that he get neuropsychological testing, which would put his client through a slew of cognitive and personality tests. That was the gold standard for identifying potential information-processing deficits. Summer said that they had done all that and all the results had been normal.

I waited for the inevitable shoe to drop. It only took a beat. Summer wanted to know if we could use fMRI to determine whether something was wrong with his client's brain. Something that the neuropsych testing didn't reveal.

On the surface, this may seem plausible. After all, various forms of brain imaging have been used in the legal arena for decades, especially since the University of Texas "Tower Sniper" of 1966. Charles Whitman, who killed seventeen people from the tower on the UT campus in Austin, and who was later killed in a shootout, was discovered at his autopsy to have a small brain tumor. The governor of Texas at the time, John Connally, put together a commission of experts to determine what, if anything, the tumor had to do with the shooting. In the end, the commission couldn't determine whether the tumor had caused Whitman's behavior, but because of its location next to the amygdala, the commission thought that the tumor might have contributed to loss of control of emotions and actions. Ever since, astute defense attorneys have looked into their clients' brains for such abnormalities. The practice raises interesting questions of culpability and the dubious defense of "my brain made me do it."

Nowadays, it has become standard practice in all capital murder cases to obtain an MRI. Whether judges and juries are swayed by such evidence is not at all clear. My late colleague at Emory, Scott Lilienfeld, had argued that presenting neuroscience data biased juries because brain images look more scientific than psychology questionnaires. But none of this was standard in 2002. In fact, I didn't know any colleagues who had been retained as an expert witness in a murder case. Naturally, I jumped at the chance to scan the brain of an alleged murderer.

I have never written about what transpired, and only mentioned the case to my closest friends. In revisiting my notes, I intended to reach out to Summer to ask what he thought after having twenty years to reflect on what we did. I learned that sadly, Summer had passed away in 2016 from ALS. He was only fifty-six.

FROM THE OUTSET, I TOLD SUMMER THAT I WOULD NOT HELP him with an insanity defense. In 2002, there was no indication that MRI or fMRI could provide any reliable information about whether a person was sane (sanity is a legal determination, not a scientific one). Summer assured me that that was not his goal. He was seeking mitigating evidence to save his client from execution. I could get behind this. I did not support the death penalty (nor do I now).

I didn't know it at the time, but Summer had become a sort of whiz kid in legal circles for his creative applications of biology, particularly in cases involving the death penalty. In 1995, Summer argued for genetic testing on behalf of Stephen Mobley, who had murdered John Collins in 1991 during a robbery of a Domino's Pizza. Mobley had confessed to shooting Collins in the back of the head while he begged for his life.[2] Summer's job as court-appointed attorney was to save Mobley from the chair by presenting mitigating evidence. There wasn't much. Mobley was white and twenty-five at the time of the crime. His family was relatively affluent. He hadn't been abused. Even so, Mobley had been in trouble most of his life, cheating, lying, stealing, and, eventually, committing armed robbery. While awaiting trial for Collins's murder, Mobley had "Domino" tattooed on his back.

Summer discovered that the Mobley family was known among the locals to produce both reprobates and successful businessmen. Summer latched onto a recent article about a genetic mutation in the gene for monoamine oxidase A (MAO-A) associated with violent behavior in a Dutch family.[3] Summer thought his client might have the same abnormality. If he did, that might provide enough mitigating evidence to avoid the death penalty. One of the authors of the Dutch study offered to help. Yet the State Supreme Court did not buy Summer's argument and denied his request for genetic testing.[4] Mobley was executed in 2005.

I was not aware of any of this when I met Summer. Would it have changed anything? Probably not, but in the case of Alan Jones (not his real name), Summer was executing a plan similar to the one he'd designed for Mobley. Instead of genetics, though, he was interested in brain imaging results. And, I suspect, having learned his lesson in the Mobley case, Summer wasn't going to ask the court for permission to do the testing. He was just going to do it and let the court decide what to do with the results.

The facts of Jones's case bore an eerie similarity to Mobley's. On a dreary December evening, Jones and his two friends broke into a neighbor's home. They were planning a burglary and didn't expect the owner to be home. But he was, and Jones shot him dead. Realizing what he had done, Jones tried to obscure the crime scene by setting fire to the body and the house.

When I met Summer for the first time, he threw down a stack of crime scene Polaroids on my desk. I caught a glimpse of charred remains before I admonished him for biasing me. He apologized and scooped them up. Summer had the idea of showing the pictures to Jones while we scanned his brain. He wanted to see whether Jones's brain reacted abnormally to seeing the crime scene. I pointed out that there was no standard for how someone should react to seeing such pictures. Would a failure to react indicate that Jones had been at the scene? The alternative—a strong emotional response as evidenced in the amygdala and hippocampus—might indicate recognition, or it could just as well reflect shock and revulsion in a normal person. There was no science for this.

In 2002, there were no instances of fMRI being used in a trial, so we would be breaking new legal ground. I'll admit, this was a big part of the attraction of the project. I was looking forward to testing my ability as an expert witness. But if there is no scientific basis for a finding, it is inadmissible in a court of law. There is substantial case law around the admissibility of evidence, and two rulings in particular dictated how we should approach the use of fMRI. The first standard addresses the admissibility of scientific evidence itself. In *Frye v. United States* (1923), the court dealt with an early form of a lie detector. The expert argued that blood pressure could be used as a measure of deception. This was a novel idea at the time, and the court disallowed it, ruling that scientific evidence should be admitted only if it has gained general acceptance by practitioners in the field. In part because of *Frye*, lie detectors and polygraphs have never been admitted as evidence, because most experts do not consider them as reliable arbiters of truth and lie telling. In modern practice, the *Frye* standard is interpreted to mean that the basis for the scientific evidence must have, at a minimum, been published in a peer-reviewed journal. That still may not be sufficient to show that it is generally accepted, but it meant that whatever Summer and I did, it would have to be drawn from the published literature. Brain reactions to crime scene photos definitely did not meet the *Frye* standard.

In 1993, the Supreme Court had broadened the *Frye* standard from the evidence itself to the expert's opinion of it. In *Daubert v. Merrell Dow Pharmaceuticals*, the court provided a number of criteria for judges to consider. For example, scientific evidence must follow from the sound application of the scientific methodology. In other words, DNA is not admissible if the expert doesn't take precautions to prevent cross-contamination. The criteria, which were added to the Federal Rules of Evidence, are periodically revised for clarity. In 2002, *Daubert* took precedence and could be boiled down to these three rules: (1) Testimony must be based on data. (2) Testimony must be the product of reliable principles and methods. (3) The witness must apply the principles and methods properly.

We holed up in a conference room and put the three *Daubert* criteria on the board. The first would not be a problem. fMRI was, in 2002, well established as a technique that measured correlates of neural activity. The kicker was number 2. We needed an fMRI test that was established as measuring a specific cognitive process and that was plausibly deranged in Jones's brain. Summer thought his client had some information-processing deficit that affected his judgment on the fateful night. But bringing a gun to a burglary indicated forethought and planning, so he couldn't argue that it was purely an impulsive act.

I WAS HAVING A MAJOR CLASH OF SACRED VALUES. IT WAS hard to see it from Summer's point of view. Jones's role in the crime really wasn't in doubt. He had taken the life of an innocent person. As a consequence, the state was preparing to take his life. It was the classic "eye for an eye." But these sacred values—*thou shalt not kill* and *an eye for an eye*—conflict with each other. In narrative terms, each contains a millennium of stories, no doubt too complicated to explain, hence the compression down to simple rules.

There was no easy resolution to these conflicting narratives. Moreover, I had stepped into Summer's legal world, which was a place very different from that of science. I was accustomed to shades of gray and multiple interpretations for scientific findings. A legal finding, though, is black and white. You are either guilty or innocent. In my mind, I never bought into the equivalency of retribution. Killing someone was wrong, and killing the murderer didn't make sense in the karmic algebra. I reasoned that if there really was something objectively wrong with Jones's brain, well, that wasn't totally his fault.

We made a list of cognitive processes that were likely to be active in Jones's brain at the time of the crime. Fear, obviously. Jones's fear of getting caught, but also his victim's fear. Did Jones not recognize the terror in the victim's face? Psychiatrists and psychologists had a term for this: alexithymia, or emotional blindness, which was the inability to identify emotions experienced either by oneself or by others. Fortunately, there was a wealth of published results on the brain structures associated with the recognition of emotions, especially in facial expressions.

The modern science of facial expressions originated with Charles Darwin, who believed that the expression of emotions in humans had biological origins that could also be found in animals.[5] With evolutionary origins, emotions were thought to be universal in humans. Although emotions are internal and subjective, their expressions are not. The reason humans are so outwardly expressive continues to be debated, but the fact that we can recognize expressions of joy, fear, and disgust irrespective of culture comports with Darwin's theory.[6] In the 1960s and 1970s, Paul Ekman, a professor of psychiatry at the University of California, San Francisco, blazed a path in the study of the expression of facial emotions. He created a standardized set of eighteen black-and-white headshots of actors portraying different emotions, which he then showed to people of all ages and cultures. Across ten cultures, ranging from the United States to Greece, Italy, Japan, and Sumatra, his team found a high level of agreement for the recognition of happiness, surprise, sadness, fear, disgust, and anger.[7]

By 2002, researchers had taken Ekman's faces into the MRI. Two critical findings had emerged. First, circumscribed portions of the visual system were highly selective for faces.[8] These face areas, located toward the rear of the brain just downstream from where visual information first hits the cortex, responded preferentially to faces over objects or landscapes. These face areas seemed to be involved in low-level processing, extracting the general configuration that a face was present. Abstract faces, consisting of no more than three solid circles arranged in an inverted triangle, were often sufficient to evoke responses in face areas. Second, the emotional content of facial expressions seemed to be processed in a different set of regions. Researchers had begun to home in on the amygdala—a structure known to be involved in arousal and memory. Pictures of angry or fearful faces elicited activity there, while neutral and happy faces did not. A recent paper had suggested that the amygdala response could be elicited even by subliminal presentations of faces. Here, angry faces could be flashed for thirty milliseconds—too quick to consciously register, but not so fast that the amygdala didn't notice.[9]

Given the published track record of these face paradigms, both inside and outside the MRI scanner, they met the *Daubert* standard of reliability and acceptance. We decided to show Summer's client pictures of faces with different emotions. We would do this both at a normal presentation rate, so that we could get a baseline of Jones's facial processing system, and we would also do it subliminally to isolate the amygdala response from any conscious attempts to show an expected response. This was critical because, we reasoned, an opposing expert might argue that anyone could fake an expected reaction to an angry or fearful face, but not to one flashed so quickly that they didn't even know it was there.

It's possible, of course, that Jones could read faces but had poor impulse control. So we planned for a different type of probe as well. Psychologists talk about impulse control in different ways, but the core idea has to do with inhibiting an action that a person wants to do but knows they shouldn't. It can be as simple as a game of Simon Says, in which a person is lured into following directions until the key phrase ("Simon says") is left out, but the person still moves out of habit. On the opposite extreme is the legal defense of uncontrollable impulse, which means an impulse is so irresistible that it cannot be avoided. It falls short of most tests of insanity, which require the defendant to be unable to tell the difference between right and wrong. Instead, the irresistible impulse is often used as proof of diminished capacity, a sort of mitigating evidence that is applied to petition for a less severe penalty. However, the lawyers for Lorena Bobbitt successfully argued for an irresistible impulse to cut off her husband's penis after years of abuse. She was found not guilty.

Summer thought something similar might have been operating in Jones's brain. It came down to the MRI to show that Jones had some biological predisposition to poor impulse control. It was already well established that the frontal lobes were critical for this function. The case of Phineas Gage, who had part of his frontal lobes blown out by a railroad iron in 1848, is the classic example showing the importance of these brain regions for impulse control. Jones, though, would not have anything so obviously wrong with his brain. At best, he would have a functional deficit, meaning the gross structure would look normal but wouldn't function properly under stress.

This might seem like a fishing expedition, but there were scant data to go on. One of the only published neuroscience studies of murderers had, in fact, shown diminished activity in their prefrontal lobes.[10] The easiest way to probe impulse control in the prefrontal lobes was to use the "Go/No-Go" task. The task presents a sort of visual countdown with a descending series of asterisks on the screen, starting with five asterisks and counting down to one. This was followed by either the letter X or A. All Jones had to do was press a button for every X, but not for A. In the vein of Simon Says, the task is set up to have more Xs than As, so the subject gets used to hitting the button. Then, when an A appears, he has to put on the brakes and inhibit the impulse to respond. When neurotypical people do this, a portion of the prefrontal lobe called the anterior cingulate cortex (ACC) becomes active as the person successfully inhibits their response.[11] If Jones had diminished impulse control, we reasoned that it might show up as a diminished response in the ACC.

We had our plans. The only thing left to do was to put the man in the scanner.

JONES WAS NOT AT ALL WHAT I EXPECTED. WERE IT NOT FOR his clothes and his escorts, he could have been any undergrad. He spoke to me with a level of respect I was unaccustomed to, ending every sentence with "Sir." In fact, he was overly compliant and seemed a bit in awe of the technology. How could I jibe this with the narrative of a cold-blooded killer who had set fire to his victim's corpse? I couldn't.

I asked the guards to unshackle Jones for the MRI. They didn't seem to mind, probably because they had already decided he wasn't a flight risk. I did, however, have to stop one of the deputies from following him into the magnet room. I pointed to the gun on his belt and told him that no metal was allowed. He hesitated. Don't worry, I said, there are no other ways out of the room. He remained in the control room with our team, while the other stationed himself outside the MRI suite to make sure nobody entered or exited.

The MRI itself was uneventful. The initial structural scans appeared normal. No asymmetries or tumors, à la Charles Whitman. For the face tasks, Jones didn't have to do anything. He just watched what appeared on-screen while we collected brain responses. Even the Go/No-Go task, which required his active attention, went smoothly. If there was anything wrong with his brain, this task would be the most likely to pick up any abnormalities. A decade later, the first large-scale study of inmates would show diminished responses in the ACC of criminals, and the more blunted the response, the more likely an inmate would be rearrested within four years.[12]

On the conscious face task, we found robust responses in the fusiform face area (FFA), indicating that Jones was paying attention and that the basic visual circuitry was intact enough that his brain could perceive faces. In other words, he didn't have prosopagnosia—aka face blindness—which results from damage to the FFA. This was not surprising. But his response on the subliminal task was. His amygdala unambiguously responded more to pictures of fearful faces than to happy ones, just as was reported in the literature. In fact, the magnitude of the response was almost identical to the reported average of a group of normal subjects. The Go/No-Go results were less informative. The ACC showed a negative response for both Go and No-Go trials, which was opposite to what was reported in the literature. Other target regions in the prefrontal cortex did show positive responses and were greater for the No-Go trials. The fact that Jones had very few errors on the task was consistent with the higher activity on the No-Go trials, indicating that he could inhibit a motor response when called to do so.

I called Summer with the results. We hadn't found anything that would help his client. Jones had a perfectly normal, average brain. We could find nothing wrong with it, either structurally or functionally. Summer thanked me for the effort, and that was that. Even though our brain scans had nothing to do with it, Summer ultimately succeeded in saving Jones from execution. Jones is serving a life sentence without the possibility of parole. He is incarcerated in a remote corner of Georgia in a facility known for housing the most violent criminals. It is frequently cited for having inoperable locks and inmate-on-inmate violence.

NEUROSCIENCE HAS HAD A CHOPPY RELATIONSHIP WITH THE law. It has become common to introduce brain scans in capital murder cases, not so much because there is a consensus as to their utility but because defense attorneys and judges want to avoid a mistrial if it is later found that such evidence might have been useful. The utter normality of Jones's brain echoed Hannah Arendt's concept of the banality of evil.[13] Jones's brain was not detectably different from that of any of the undergrads who volunteered for our experiments.

The banality of the brain cautions us against looking too hard for biological explanations for things that people do. I have no doubt that the events of that December night in 1999 are still housed in Jones's brain. But, as we have seen over and over, the brain's storage medium is highly compressed. The task of extracting what happened, and what was in the mind of a killer—mens rea—is akin to reconstructing a movie from a synopsis.

Maybe first-person autobiographical memories are stored in more detail than third-person ones. Even so, they still need to be decompressed before there is any chance of measuring them through technological means like MRI. We did not ask Jones to remember the events leading up to his crime. There was no point. We didn't have a template of what that would look like in the brain, and we still don't. The technology has improved in the twenty years since, mainly in the application of AI to big-data problems like brain decoding, and we may yet unravel the basis function for how murder is stored in the brain. I suspect, though, that such memories would have to be disentangled from the personal narrative in which they are embedded.

At the outset, I had hoped that there was going to be something remarkable about Jones's brain. My hope was driven by the desire to help save a man from execution but also the idea baked into my own moral compass that people who commit murder are somehow different from the rest of society's upstanding members. The uncomfortable reality, though, is that we are all more alike than different. Our inability to extract something as profoundly awful as killing another human being points to the banality of the brain. It is not the physical structure of the brain but the contents within it that determine one's life story. Granted, we weren't trying to unravel the events of that night in 1999, but the normality of Jones's face processing and impulse control told us that if there is something different about a murderer's brain, then it must be subtle indeed. Murderers are individuals, too, and it is unreasonable to expect that all killers are the same. Some, like Jones, could be basically normal people but ones who make bad decisions. Based on what Atran and I had discovered about sacred values, I am tempted to conclude that in the case of Jones the problem lay in a breakdown of deontological resolve.

The flip side of this explanation implies that the capability to kill is more widespread than we might want to acknowledge. I have touched on some of the brakes that keep such impulses in check: the Golden Rule, Ten Commandments, theory of mind, and empathy. Different societies have different rules of acceptable conduct. Currently, most do not sanction killing, except under certain circumstances, like war or some types of police encounters. We return, then, to the idea that much of what we think of as the core of personal identity is, in fact, simply the rules of the particular society we live in. Maybe the self delusion is also a mass delusion. In the next chapter, we'll see how society shapes not only what is acceptable to do but also what is acceptable to think.

CHAPTER 13

The Man with Half a Brain

T HIS SECTION HAS DISSECTED THE CONSTRUCTION OF shared narratives, or what I would call the consensus version of reality. We've examined how other people's narratives worm their way into our brains, constantly nudging our personal narratives to comport with a version that is deemed acceptable. You might call it the conformity of narrative, or less generously, the reality of a shared delusion. Because the brain stores memories in a compressed format, information is necessarily thrown out. What's left is an approximate representation—a fiction—of an event that happened. Moreover, because one person will compress an event differently from how someone else does, the memories of two people who both experienced the same event can vary widely. This becomes apparent when we tell each other stories. A husband recounts something that happened years before, and his wife corrects him with what really happened. Assuming they have a healthy relationship that can weather disagreements, they will kick it back and forth and come to a shared version of events.

This is how we define reality, through a shared understanding of the world.

Objective measurements like those obtained from cameras and scientific instruments don't change the facts. The interpretation of their outputs must still be filtered through each individual's brain. This is why police body cam video is not always helpful. There are still multiple interpretations of what was recorded.

We are now faced with the unsettling conclusion that no thought is truly our own. Even our own memories are suspect. They are highlight reels, prone to confabulation to fill in the holes. More often than not, these holes are filled in with things that we have read or heard from other people. Is reality, then, just a shared delusion? To some degree, yes. But the distinction between a useful delusion, such as the one that binds one's identity together, and something more extreme is not always clear. A large portion of the population harbors versions of a personal narrative that they keep to themselves. You might say that they are telling stories that take them on journeys less traveled. And although they might be extreme, there is something intriguing about the fluid nature of how their brains craft a personal narrative. They might just hold clues for changing who you think you are.

DELUSIONS ARE THE HALLMARK OF SERIOUSLY DISTURBED brain function. When they are the result of an identifiable medical process they are called "organic." Drugs—especially sedatives, anesthetics, and opioids—have well-known tendencies to cause odd beliefs and sometimes hallucinations. (Psychedelics, such as LSD and psilocybin, alter perception but do not generally cause delusions.) Certain medical conditions can cause delusions. When the brain degenerates to a sufficient degree, delusions become common in Alzheimer's and Parkinson's patients. Autoimmune diseases like lupus can cause delusions when the brain is inflamed.

Sometimes there is no medically identifiable cause. When all the organic causes have been eliminated, delusions are said to be functional, that is, psychiatric. Such delusions tend to occur in three states: depression, mania, and schizophrenia. Indeed, the history of psychiatry can be traced by its relationship to the disease we call schizophrenia, a Latin term that means "splitting of the mind." In modern usage, schizophrenia refers to a set of mental symptoms that manifests as a break with reality. A stereotypical picture of schizophrenia looks like a homeless person who is disheveled, smells like urine, talks to themselves, and rails against invisible demons like the CIA. As with all illnesses, however, there is a tremendous range of expression of symptoms and severity. It is true that patients with schizophrenia can sometimes have a tenuous relationship to reality. When the symptoms are at their worst, the sufferer operates under a personal narrative that seems wholly foreign to others. Schizophrenia shows us how personal narratives are defined by a shared reality, or what we would call culture. It is precisely when the personal narrative goes off the rails that we see clearly that nobody's narrative is truly their own. When someone's narrative deviates too far from the norm, they are deemed crazy.

The thing about delusions is that they are maddeningly difficult to separate from their owners. First-year psychiatry residents learn this the hard way. They all go through a phase of treating patients with delusions—usually longtime sufferers of schizophrenia—by performing reality checks against a range of odd beliefs. Without medication, confronting patients with facts rarely breaks their belief system.

In my second year of residency at Western Psychiatric Institute and Clinic in Pittsburgh, I treated a patient named Barry (not his real name). WPIC was a leader in psychiatric research and treatment. The building itself towered over the Oakland section of Pittsburgh, bathing everything beneath it in the warm glow of its orange brick facade. Each of the thirteen floors housed a different

type of psychiatric unit. On the first floor, the psychiatric emer-
gency room was located just off the main lobby. There was a pedi-
atric floor. There was a geriatrics floor. There was a mood disorders
unit. The top floor was for eating disorders.

And there was the schizophrenia unit. Technically, it was the
general adult unit, but because psychotic patients can be dangerous to themselves and others, they were housed together so that
their particular needs could be efficiently handled in one location.
Most of the patients carried some variant of schizophrenic diagnosis. Labeling someone as schizophrenic is a big deal. Schizophrenia
is one of those awful diseases for which there is no cure. It is a
diagnosis that will be attached to a person for the rest of their life.
And the prognosis is not good. Although less than 1 percent of
the population suffers from schizophrenia, because it is a chronic
illness that often prevents a person from steady employment, it
carries a high economic burden for its sufferers and society's safety
net. People with schizophrenia die by suicide at a much higher
rate, and they also suffer from medical conditions at roughly three
times the rate of the rest of the population. Because their condition interferes with getting proper care for heart disease, diabetes,
and cancer, schizophrenics tend to die young.

A serious disease for sure, on top of which the label itself carries
such stigma that we wanted to be sure that a person fully met the
criteria before even bringing up the S-word. The *Diagnostic and
Statistical Manual*—aka *DSM*—was our bible. It was by no means
an ideal approach to diagnosis, and a patient had to have a specified number of symptoms to garner a diagnosis. When I was in
training, the *DSM* was transitioning from the third to the fourth
edition. For schizophrenia, a person had to have two or more of
these symptom clusters for at least a month: delusions, hallucinations, incoherent speech, catatonic or disorganized behavior, and
so-called negative symptoms that included a decrease in speech
or flattening of emotion. If the delusions were really bizarre or the

hallucinations consisted of an internal voice keeping a running commentary, then those alone would suffice. In addition to the one-month duration, it was required there be at least six months of lesser symptoms, which are called prodromal because they precede the development of full-blown psychosis. Because drugs can also cause these symptoms, the person should be free of drugs before labeling them as schizophrenic.

Although the symptoms may seem pretty obvious, the duration was rarely so. And the duration was key. It should surprise no one that psychotic patients tend to be poor historians. They have a tenuous relationship to reality, and the timeline of events in their lives can get jumbled.

Barry carried the diagnosis of Psychosis Not Otherwise Specified (NOS). That meant he had some of the symptoms of schizophrenia but hadn't met the duration requirement. He was only twenty years old, which was about the typical age for a first-break psychotic episode in schizophrenia, but nobody on the treatment team was ready to say that that was what he had.

He was quiet and kept to himself on the ward, but unlike many of the other patients, Barry was never disheveled. He was always clean-shaven and wore his black hair cropped close, almost military style. He gave a sly smile when I introduced myself. He shrugged when I asked him why he was in the hospital.

I knew his history but wanted to hear what Barry thought of it. He had been a sophomore at a prestigious college. He returned home after a semester of disastrous grades. According to his chart, Barry had become consumed with a specific delusion. He was convinced that he had only half a brain. Over time, he could think of nothing else and stopped going to classes. He did not hear voices or suffer from any of the other schizophrenia symptoms. He didn't do drugs, and his blood and urine were clean as a whistle.

I asked Barry what had happened to his brain.

It shrunk.

How did he know?

He felt it.

If you only had half a brain, I said, you would be paralyzed on one side of your body. This was a tactical error on my part. I should not have been challenging his delusion without first establishing some type of rapport.

Barry shook his head. There are plenty of people walking around with half a brain, he said. The remaining half can take over.

Technically that was true, but that was a phenomenon usually seen in children.

Barry was scheduled for an MRI. I guess we'll see, I said.

He seemed indifferent and just nodded.

I was flummoxed. In our brief interaction, I sensed there was no way to crack open Barry's delusion. I knew what the MRI would show. There was no way he had lost half his brain. Not without some major impairments.

IN CHAPTER 7, WE SAW HOW SUPERSTITIONS CAN BECOME so powerful that they guide a person's behavior. Is there anything fundamentally different, then, between a superstition and a belief in half a brain? Both dictate what a person does. It just happened to be that in Barry's case, it resulted in him dropping out of school. Psychiatrists call that a significant impairment, and it is arguably why he was in the hospital. Superstitions are usually not incapacitating. The DSM placed special emphasis on bizarre delusions, such that if a delusion was sufficiently strange, no other symptoms would be necessary for a diagnosis of schizophrenia.

But what counts as bizarre is itself subjective. Psychiatrists continue to grapple with this conundrum. There are two possibilities. Either delusions are fundamentally different from the garden-variety type of odd belief that most of the population holds; or delusions exist on one end of a spectrum of beliefs, perhaps extreme in strangeness or imperviousness to argument. The debate has practical implications for how schizophrenia is treated. It also tells us something important about the nature of personal narratives, which, as I have argued, can be considered a form of delusion.

Let's consider the first theory, that psychotic delusions are categorically different from everyday odd beliefs, because that is the more widely accepted view of psychosis. This explanation is linked to the early theories of schizophrenia. In the late 1800s, psychiatrists recognized that there were two forms of psychosis. One form seemed to wax and wane, with the patient often recovering completely between episodes. Patients with the other form never got better. It was Emil Kraepelin, a German psychiatrist, who first put labels on them. The first he called manic depression (what we now call bipolar disorder). The second form he called dementia praecox. In 1908, the Swiss psychiatrist Eugen Bleuler replaced the latter term with schizophrenia.

The early 1900s were a period of intense development in psychiatry. Physicians were not only trying to treat patients with mental illness but also vigorously debating the phenomenology of the diseases themselves. Although Kraepelin had correctly identified the difference between bipolar disorder and schizophrenia, in practice, it wasn't so easy to tell the difference. Given that one disorder tended to have periods of normality, often the only option was to institutionalize psychotic patients. If they eventually got better, they could return to their lives. For those who didn't, well, they might never leave.

By 1920, a group of psychiatrists in Munich were cataloging the nature of schizophrenic symptoms. Karl Jaspers and Kurt Schneider aimed to increase diagnostic precision by this approach, which would later be adopted in the *DSM*s. When it came to psychosis, they thought a delusion should be defined by the way a belief was held, not by its content. A belief that grew out of a normal perception of something considered commonplace that morphed

into something of special significance was particularly worrisome
for schizophrenia. Schneider later included this type of delusion
in his list of first-rank symptoms of schizophrenia. Other first-rank
symptoms included auditory hallucinations taking the form of
voices in a person's head, especially those berating the individual
or providing a running commentary. First-rank symptoms also included the belief that a person's thoughts were the result of some
external force (called thought insertion) or, conversely, that a person's thoughts were accessible to others (thought broadcasting).

Jaspers and Schneider's phenomenology is still used today, although first-rank symptoms are now known to be common in bipolar disorder. Critically, the twentieth-century psychiatrists did
not include the content of delusions in their system. My patient
Barry's belief that he had only half a brain was precisely the sort
of first-rank delusion that worried me. I suspected it probably did
originate in a garden-variety perception, like a migraine headache,
which frequently can be one-sided, and then blossomed into the
improbable conclusion that half his brain had disappeared. In Jaspers's view, the delusion didn't result from a disorder of perception.
Rather, it resulted from a transformation of meaning.[1] If it was a
matter of perception, then a person might be dissuaded from their
belief by presenting counterfactual evidence.

Of course, Barry's MRI came back normal. And, as expected,
the result didn't budge his delusion. Only time and medication
shifted his belief. We started a new antipsychotic called risperidone. Like most other antipsychotics, risperidone blocked the receptors in the brain for the neurotransmitter dopamine. Why this
works to quell psychotic symptoms remains a mystery, but ever
since the discovery that Thorazine—also a dopamine blocker—
could treat schizophrenia, the conceptualization of the disease
has centered on dopamine. The two are so closely linked that
the *dopamine hypothesis of schizophrenia* remains very much alive,
albeit in modified form.[2] For decades, it was thought that excess

dopamine caused the positive symptoms of psychosis that Jaspers and Schneider had identified. This was based solely on the observation that dopamine-blocking drugs improved these symptoms.

But repeated neuroimaging probes and genetic assays have not offered proof for this theory. Schizophrenia is clearly a brain disease distributed through many systems. Other neurotransmitter systems implicated in psychotic symptoms include the N-methyl-d-aspartate (NMDA) receptor, which can be blocked by the drugs phencyclidine (PCP) and ketamine and which often causes hallucinations and paranoid delusions. Similarly, the serotonin system plays a key role because this is where psychedelic drugs like LSD and psilocybin act. There is growing evidence for the therapeutic uses of psychedelic drugs and their potential for catalyzing psychic change in the individual.

In Barry's case, the blockade of his dopamine system did not cause an instantaneous change in his delusion. Over a period of weeks, he gradually stopped talking about it. He still thought he had half a brain. I know this because I asked him every day. With time, though, the belief receded into the background and was not the primary motivator of his actions. By the time he was ready to go home, Barry passed as normal in casual conversation. I couldn't say whether his odd belief would manifest in five minutes of talking to him, or an hour, or whether it had receded enough that he had gained enough insight and self-control to keep it to himself. As I hugged him goodbye, I worried whether he would be able to return to college and, if he did, whether the stress would cause his delusion to take control of his life again.

BARRY'S STORY MAKES A CASE FOR THE SECOND WAY OF thinking about psychosis, that delusions exist along a continuum of beliefs. At face value, the belief itself is no stranger than other things that people believe. Sister Boniface Dyrda believed that the ghost of Father Junipero Serra cured her illness. Not only that, she convinced leaders of the Catholic Church. When I was in medical school in San Diego in the 1990s, I lived down the street from a nondescript ranch house that was widely known to be the base of a cult. I didn't learn their name—Heaven's Gate—until after they committed mass suicide in an effort to reach an alien space-ship they believed was hidden behind Comet Hale-Bopp. Former members of the cult still maintain a website.[3] Were the thirty-nine people who died psychotic? If they were, then we would have to say that all the members of the Church of Scientology are psychotic, too. Their belief system is even odder than that of Heaven's Gate.

As with superstitious beliefs, sporadic psychotic symptoms are found to be fairly prevalent in random samples of the population. The US National Comorbidity Survey, which was conducted in 5,877 people in 1990–1992 and again in 2001–2002, aimed to measure the prevalence of a range of mental health symptoms. Approximately 28 percent of adults had at least one psychotic screening question. The most common symptom, *Have you ever believed that people were spying on you or following you?* was endorsed by 12.9 percent of people. The next most common was: *Have you ever had the experience of seeing/hearing something that others could not?* When clinicians followed up with these people, the preva-lence of diagnosable psychotic illness fell to approximately 1 per-cent.[4] More detailed analysis showed that individuals could be grouped along a continuum of severity by roughly the number of psychotic-like symptoms they endorsed.[5] Similarly, when such sur-veys are used as the basis of diagnosis, people with intense spiritual or religious beliefs are often indistinguishable from hospitalized psychotic patients.[6]

The range of psychotic symptoms in the general population comports with the observed effect of medications to modulate such beliefs. Dopamine blockers turn them down, while NMDA and serotonin-specific drugs can crank them up. It means that delusional tendencies live on a continuum.[7] There are two broad implications of such a view. First, for patients diagnosed with schizophrenia or one of its diagnostic cousins, it may not mean a lifetime of that categorical label. Rather, like many chronic illnesses, schizophrenia can wax and wane, with periods of relative normality. Of course, this would require diligent management through treatment and medication, just like heart disease or diabetes. Second, a continuum of psychosis means that a significant chunk of the population—at least 28 percent—lives in the penumbra of bizarre beliefs.

The state of affairs leaves us to reason that one's relationship to reality is, at least externally, defined by the people around you (and their belief in your sanity). The ability to describe and converse about shared events in the world requires a certain level of agreed-upon ground rules. Mathematicians call them axioms—truths that we hold to be self-evident but that cannot be proven. Early on, we saw how first stories lay these ground rules, or basis functions.

We cannot deny the effect others have on one's perception of the world, but the world includes not only others, but oneself. Does anyone dare share their true perception of themselves? Not fully. If we voiced what was in our heads all the time, we would surely consider each other insane. Look deeply inside and you may see an epic story of such grandiosity that a psychiatrist might label you as manic or evincing extreme narcissism. Perhaps a picture of self-loathing lies beneath. Or, more likely, maybe both exist in eternal conflict. It is only by the skin of our frontal lobes that we keep these perceptions to ourselves. That is why dopamine-blocking drugs work. They help keep the genie/demon in the bottle. Nobody cares if you believe that you are the second coming of God. Just as long as you keep the thought to yourself.

This realization can be liberating. In Part III, I will turn to ways in which we can change our personal narrative. If you accept that these narratives themselves are fictions, a large palette of alternative stories opens up to you. Patients with schizophrenia show us the range of stories that people tell to themselves. Of course, society does not always accept them, but that is a risk when one departs from the views of the majority. The individual who strikes out on their own does so at their own peril, but with that risk comes the possibility of reward. For the individual, the benefit might be as simple as the satisfaction that comes with following your passions, no matter what other people think. For an entrepreneur, this might even translate into financial success, although that would require convincing a large number of other people to buy into your vision.[8] Society as a whole benefits, too. We end up with a society of diverse individuals who see the world in different ways. The alternative—enforced conformity of thought—is a dead end without possibility of innovation. In the spirit of diversification, let's see how to break free from some of the templates and basis functions that have resided in your brain since childhood and begin crafting a new self.

PART III

THE FUTURE OF YOU

CHAPTER 14

Novels on the Brain

A great book should leave you with many experiences, and slightly exhausted at the end. You live several lives while reading.

—WILLIAM STYRON,
CONVERSATIONS WITH WILLIAM STYRON

I'VE BEEN BUILDING A CASE THAT THE SELF AS A CONTIN-uous, consistent narrative is a fiction. Less generously, the self is delusion. While the specific details may vary from person to person, the templates are largely the same, filled out by details that very likely entered our brains from external sources. Now, you might object to this characterization as extreme. After all, it is not the case that our personal narratives are wholly fictional. Events happened that others can agree upon. Perhaps it's more accurate to say that our personal narratives are like historical fiction, teth-ered to events in the real world.

And like a protagonist in a historical novel, we're presented with choices throughout the narrative that can shape our destiny. Up to this point, I have painted a somewhat passive picture of how other people's opinions get into our heads. In truth, we have some degree of control over whom we listen to, the books we read, and the media that we view. Informationally speaking, we are what we eat. In this final section, we will explore the extent to which it is possible to actively change the self by choosing what we consume. Just as stories shape past-you, stories can change future-you.

To start we need to understand how, exactly, stories change the brain. As neuroscience has pushed deeper into the human mind, it has sought out explanations for constructs that used to be considered sociological, meaning the stuff of culture that binds people together. Until recently culture was thought to be a property of groups of people, but individuals make up a culture, and so, the thinking goes, there must be common traces of culture in our heads and ultimately in our brains. Stories make up many of these traces.

There is an implicit assumption here, that everything an individual encounters, whether it is in real life, online, or in a book, alters the brain. This claim has become nearly a cliché, and yet when it comes to reading, most people can identify a book (often one they read in adolescence) that changed the way they think. Stephen King has cited William Golding's Lord of the Flies, which he encountered when he was twelve years old. He described it as, "The first book with hands—strong ones that reached out of the pages and seized me by the throat."[1] We don't even have to take his word for it. Elements of Lord of the Flies permeate King's books, which reveal the innate potential in kids for both good and evil.

A book's impact is the result of an interaction between the material itself and the mental state of the reader. There are the technical facts: most adults can read about two hundred to three hundred words per minute, which means that a book of average length—say eighty thousand words—will take anywhere from four to seven hours to read. If you're reading this, then it's probably safe to assume that you enjoy reading (unless someone has assigned it to you, in which case, I'm sorry). Either way, you have probably read, or listened to, hundreds, maybe thousands, of books. If you're like me, you are probably hard-pressed to remember the details in most of them. Yet each book on my shelves represents a physical artifact of both its content and the experience I had reading it. As I walk around my house and look at my books, I recall not only the contents of each book but also, more importantly, the circumstances in which I read it.

When I was twelve, I signed up for the science fiction book club. Fortunately, when boxes of books started arriving at our house, my parents didn't send them back. Instead, they let me keep them as long as I read them. That seemed like a fair bargain. I can still remember being floored by *The Foundation Trilogy* and *Dune*. If I had to pick a book that shaped the way I think, it would be one of those. Part of it was the authors' portrayal of worlds so completely different from the one I knew. But the protagonists were even more important. I imagined myself as the psychohistorian Hari Seldon, using mathematics to predict the future of the universe, or as Paul Atreides, taming a sandworm. Even now, when I look at my copy of *Dune*, I am instantly transported back to the room of my early teenage years. I am lying on my bed, trying to understand the made-up language Frank Herbert crafted. If I had read it any other time in my life, it would not have had the same impact.

If someone says a book changed their life, well, then surely a conscious acknowledgment of such importance would change the brain somehow. The question, though, is whether these changes can be detected. And, if they can, what would they mean?

When we say that something changes the brain, there are two ways that a change might manifest. The first is a transient alteration. The majority of psychological experiments are designed around this phenomenon. These transient changes are relatively easy to detect. You define a control condition and then present the subject with a stimulus that is designed to evoke a particular response. The experimenter assumes that once the stimulus is gone, the response will return to its baseline. The response could be anything measurable. It could be a keypress on a keyboard, or it could be a physiological response like a change in heart rate, skin conductance, or brain response as measured with fMRI. These types of experiments are efficient. Trials can be repeated over and over until the experimenter acquires enough data for analysis.

The second type of alteration is a long-lasting change, but these are more difficult to measure. When it comes to the brain, transient alterations represent ephemeral change, and most neuroscientists interpret them in terms of momentary information processing rather than lasting change. The visual cortex, for example, responds to changes in the visual field, but these are not thought to persist. Once a stimulus is gone, so is the brain response. When it comes to a cultural trace in the form of literature, we would really like to know whether there is some sort of permanent alteration to the structure of the brain. This requires a different type of measurement.

A general rule of biological systems is adaptation. For example, the visual system adapts to changes in overall brightness. You are hardly aware of the differences in luminosity between an outdoor scene under the noonday sun and one inside illuminated by a soft-white light bulb. Adaptation makes it far more difficult to detect changes in the brain. It was largely for that reason that not much work had been done on how a book changes the brain.

By 2011, new methods in neuroimaging had emerged that purported to measure stable patterns of activity in the brain. Prior to this, we could only measure transient changes with fMRI, on the timescale of seconds. The alternative, structural imaging, was more like a snapshot of brain anatomy, but even that was not sufficiently detailed to capture the changes that were thought to occur as a result of immersive experience. And, as I had discovered, even a murderer's brain didn't show obvious abnormalities. The new technique was a variation on fMRI called resting-state fMRI, or rs-fMRI. The idea was to have a person lying in the scanner, awake but doing nothing. If you continuously scan a resting person's brain with fMRI over a period of about ten minutes, patterns begin to emerge. The brain shows coordinated activity in disparate regions, where the measured signals fluctuate up and down in synchrony. These are called resting-state networks or, sometimes, default mode networks, because they represent the default mode of activity when a person isn't doing anything.[2]

Neuroscientists continue to debate the functional significance of the resting-state networks. One possibility is that this activity is just the background hum of the brain, like the buzzing of a beehive. It doesn't have any functional significance other than neurons, like worker bees, quietly going about their business keeping the brain and body alive. Another, intriguing possibility suggests that the resting-state networks represent the anatomy of daydreaming.[3] Proponents of the beehive theory point out that resting-state networks are present even under light sedation, when spontaneous cognition is blunted.[4] However, anyone who has had light sedation—dental work, colonoscopy—knows that light sedation is not the same as full anesthesia.

Resting state is a bit of a misnomer because the networks can be perturbed by other tasks.[5] In one experiment, students were scanned before and ninety days after studying for the standardized Law School Admission Test (LSAT).[6] Connections within the frontoparietal resting network were found to be stronger after studying, and the researchers concluded that intensive training for logic questions strengthened these patterns. Maybe the students were thinking about the LSAT during the scan. More likely, the act of studying, especially because it had occurred repeatedly over days and weeks, resulted in a physical alteration to the brain itself, and these changes persisted into quiet rest periods.

If studying for the LSAT can result in detectable changes in the resting state of the brain, what about reading a book? I wondered whether we could capture a transformative reading experience like the one Stephen King attributed to the *Lord of the Flies*. This was the question at the heart of an experiment I undertook in 2011.[7]

THE FIRST DECISION WE HAD TO MAKE WAS: WHICH BOOK? Every day for weeks, my team members gathered around the central table in the lab. Undergrads, graduate students, research specialists, and other faculty threw out ideas for their favorite books, ones that changed their lives. One person loved poetry, but no one else did, nor would our likely pool of volunteers, whom we expected to be young undergrads. Harry Potter, of course, was all the rage, and *Harry Potter and the Deathly Hallows* was the most rabidly anticipated movie that summer. But we had to assume most of our volunteers would already have Harry Potter on the brain. So that was out. We debated about the so-called classics, but here, too, we had to assume that any Emory undergrad would have been exposed to at least a few. That ruled out my favorites: *The Odyssey*, *Crime and Punishment*, *Moby-Dick*. And forget about *Dune* or *Foundation*. Those were hopelessly out of date.

Although everyone had a favorite novel, we could not agree on any single work of fiction that we thought would impact the brain of a college sophomore. Some argued for nonfiction because it portrayed real events, but the idea was dismissed when nobody could think of a nonfiction book that changed their worldview as a teen. In the end, we settled on historical fiction, a plot-driven genre grounded in true events but told with a narrative flair that made history dynamic.

We chose the 2003 book *Pompeii*, by Robert Harris.[8] I remembered how much I enjoyed it when it came out. But because it had been published eight years earlier, nobody in the lab had heard of it. Everyone was familiar with the basic story: Mount Vesuvius erupted and rained hot ash down on the Roman city of Pompeii, burying its inhabitants. Harris brings the story to life by following the story of the fictional engineer, Marcus Attilius. There is love, sex, death, and tragedy. Would this page-turner make a lasting impression on the brain of a young adult? We hoped to find out.

For our experiment we chose to study the impact of *Pompeii* on a group of young adults. These would be teens between their freshman and sophomore years in college, around eighteen or nineteen years old. At this age most people are struggling to define their personal identity. Issues of race, gender, class, purpose, and, of course, relationships weigh upon the young adult's mind. I didn't expect that *Pompeii* would be life-altering for the participants, but I hoped that it would at least be powerful enough to cause lasting changes in their brains. Hopefully, they would identify with the hero of the story and his struggle to save the woman he loved from being buried under hot ash.

To make sure the participants actually read the material, we adopted a two-pronged strategy. First, they would have to read a physical copy of the book. Although e-readers were just becoming popular, we didn't want people skipping ahead, so we purchased a paperback copy for each subject. We then tore each book into nine sections. Participants would receive one packet each day over the course of the experiment. Second, to make sure that our volunteers were actually reading the material, they would take a short quiz every day before receiving the next packet.

As for the brain scanning, the plan was to have each participant come to the MRI center every morning for nineteen consecutive days (including weekends), where they would receive a resting-state fMRI scan. The scan would last about seven and a half minutes. They wouldn't have to do anything other than rest quietly with eyes closed. After that, they would take the quiz. To assess their engagement with the material, we would also ask them to rate how excited they became while reading. No material would be assigned for the first five days of scanning, which would serve as a baseline, including any daily reading they normally did. The next nine days would be reading days, and this would be followed by another five days with no reading so that we could see if any changes persisted beyond the active reading period.

This was the most complicated experiment I had ever designed. The logistics of convincing twenty volunteers to come in every day at the same time, without fail, for nearly twenty days, and to read an entire book in addition, were daunting. As an incentive, we offered a payment of $400. But there was a catch: we would deduct $100 for every missed session. We also prescreened volunteers for their commitment to reading with the simple question, "What was the last book you read for fun?" As students, everyone would have been reading class material throughout the academic year, but we wanted only people who also found time to read for fun. To be accepted into the study, a student had to have read at least one book for fun during the preceding school year.

In the end, nineteen participants (eleven female, eight male) made it through the whole experiment. Before getting to the imaging results, we had to address the question of whether the book had had any impact at all. Did the students identify with Attilius or Corelia—the woman he loved? Did they find the story interesting or was it just historical information without any emotional resonance? One clue came from the question about excitement we posed each day. The question varied slightly based on the day's material, but went like this, "On a scale from 1 to 4, overall how excited were you by this reading?"

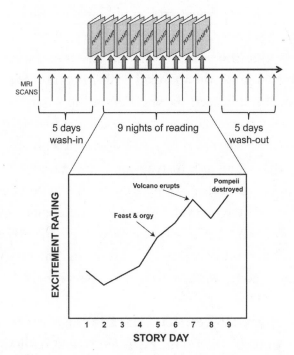

Design of experiment (above). Participants underwent resting-state fMRI scans on nineteen consecutive days. On evenings before the middle nine days of scanning, participants also read a portion of the novel, *Pompeii*. The excitement rating of each excerpt showed a rising trend toward the climax of the novel (below).

The average excitement rating began a bit above the midpoint of the scale, but started to take off on day 5, after the big Roman feast and orgy. The excitement continued to build until the eruption of Mount Vesuvius on day 7. There was a brief respite, before peaking on the final day, when Attilius and his lover are buried in the ashes, along with everyone else trapped in Pompeii. The curve looks remarkably like the canonical rags-to-riches template in Chapter 8. But because everyone dies, *Pompeii* is really a story of riches-to-rags. Although these curves are related, our question did not ask whether their excitement was good or bad. At a minimum, the excitement curve told us that the material had a measurable impact. If it had been flat, then I would have worried that the students hadn't paid attention or didn't relate to the material.

Confident that the book had caused at least a transient change in subjective feelings, we analyzed each person's resting-state activity across the nineteen days to determine where in the brain these changes occurred. Given how excited the students became, I expected changes in brain regions associated with emotions. But that is not what we found. Instead, we found a network of regions organized in a hub-and-spoke pattern, with the hub centered on an area of the left temporal lobe called the angular gyrus. This region is known to be critically involved in language comprehension. This change in connectivity represented a carryover effect of the actual reading, similar to how a muscle feels the day after exercising.

This interpretation becomes clearer in the context of the predictive functions of the brain. As I mentioned earlier, the brain is never really at rest, and the default networks can be considered a suite of modes that dynamically shift between one another as the mind turns inward, processing recent events.[9] In the case of our experiment, the recent events occurred within the novel, and the change in brain connectivity reflected the incorporation of those events into each person's brain and personal narrative. The changes in the temporal lobe were present only during the reading days. After the novel was finished, the connectivity pattern returned to its previous configuration.

To see whether the novel caused longer-lasting changes, we also looked for a different pattern: one that occurred with the onset of reading and then persisted after the novel was over. Only one network showed this pattern: the sensorimotor strip. This was a surprise, because it wasn't the emotional regions I had expected. The sensorimotor strip occupies the bank of folds along the central sulcus, where tactile impulses enter the cortex and motor impulses leave. So, why should a novel change the pattern of activity there?

One possibility is that reading the novel invoked neural activity associated with bodily sensations and that these activity traces carried over to the resting-state scan. *Pompeii* is, in fact, a visceral book. The descriptions of Roman feasts and orgies and the showering of molten ash on the inhabitants of the city can make one's skin crawl. This explanation fits with the theory of embodied semantics, which says that the brain regions responsible for producing an action are also used to represent the action in your mind.[10] In other words, when you read about someone hitting a home run, your brain unfolds a compressed representation of hitting a homer and then uses your sensorimotor cortex to simulate it. Similar results have been observed for the sensory side. In another imaging study, participants read tactile metaphors, including expressions like "hot-headed," "unbending attitude," "weight matters," and "coarse language." The simple act of reading these phrases was associated with activity in the sensory strip, suggesting that tactile concepts repurpose the same regions we use for physically feeling things.[11]

Literature immerses the reader in a world created by the writer. In many novels, the reader feels like they are in the body of the protagonist. This would explain the changes we observed in the sensorimotor network during the reading days. It is telling, though, that these changes persisted after the novel was finished. The experiment ended five days after the completion of the reading, so unfortunately we don't know just how long these changes might have lasted.

Even so, the sensorimotor changes resonate with the subjective feeling of how a book changes a person. Thinking back to the books that have changed my life—*Foundation*, *Dune*, and later, *The Teachings of Don Juan* and *The Monkey Wrench Gang*—I cannot really recall much detail about the plots. But I can vividly remember the characters—Hari Seldon and the Mule, Paul Atreides, Don Juan, Doc Sarvis. In hindsight, I must have been

drawn to their iconoclasm as they thumbed their noses at conventional social norms and forged their own journeys. They have all
been incorporated into my identity or, at least, who I think I am.

To UNDERSTAND WHY FICTIONAL CHARACTERS HAVE SO
much impact on us, we have to return to the universal monomyth.
Millennia of cultural evolution have shaped the human brain to
soak up the stories of these protagonists. When we read a book
whose central character resonates with our own narrative, their
story and our own amplify and build on each other. We become
more like the character—at least, in our minds—and that fuels
the resonance in a positive feedback loop. The changes in the sensorimotor networks suggest that we really do feel as if we become
the protagonist of the story we're reading. One might even go as
far as saying this resonance creates a sort of muscle memory that
gets reactivated in our personal narratives.

If it's just a matter of resonating with a character, then the medium doesn't matter. The same effects could be claimed for movies
and television. At some level, everything changes the brain, so
the question is one of degree. Parents and psychologists have been
wringing their hands over the harmful effects of TV on children
since broadcasting began in the 1940s. Not much has come of it.
There seems to be nothing inherently bad about television. The
main effect is its potential to displace other activities that children—and adults—would otherwise be doing.

But if we're honest about TV and movies, they do not provide
the consistent depth of engagement that a book does. First, television and movies are consumed passively. Films and shows can
be consumed independently of any effort of the viewer. With a
book, you can't consume it without effort. Second, a typical film
lasts about two hours, which is less time than it takes to read a
book. The rise of binge-watching has changed that, where a series

like *Game of Thrones* or *Breaking Bad* can take seventy hours to consume. However, reading all the books in the *Game of Thrones* series would take most people at least a hundred hours. Finally, because watching television is not cognitively demanding, it lends itself to bingeing.

When we did the *Pompeii* experiment, we purposely did not want the volunteers to binge-read. The primary reason was that we needed all the participants to maintain the same reading pace, but the other reason had to do with the way information is incorporated into the brain in a process called consolidation. During waking hours, the hippocampus holds new information in a short-term memory buffer. This material does not get incorporated into long-term stores until you go to sleep, at which point the day's information is replayed during the cycling between deep sleep and REM sleep.[12] The stuff of dreams is the shuffling and incorporation of new memories with old ones. When you read a book over days and weeks, there is plenty of time to consolidate the material. Whereas when you binge-watch (or binge-read), the hippocampus is overloaded, and less material is retained. And with less retention, there is less material to incorporate into the brain.

Video games are the only other form of media that might affect the brain strongly enough to change a personal narrative. Gaming platforms have advanced to such an extent that highly realistic and immersive environments, rendered in astonishing detail, can be experienced through devices we have at home.[13] Yet it is difficult to generalize about their impact. There are many types of video games, so nobody would expect all games to affect the brain in the same way. Playing *Tetris* is completely different from playing *Call of Duty* or *Red Dead Redemption*. The one genre that most closely resembles reading a book (and so might affect a player in a similar way) is the first-person game. In these games, the player adopts the perspective of a game character, literally walking in the shoes of the character. First-person shooter games are very popular, especially among young to middle-aged men. Driving games and flight simulators are also first-person perspective, but they are not typically plot driven like first-person shooters. First-person adventure games are probably the closest to the novel form because the protagonist has defined goals.

Although there has been much hand-wringing over the violent content of shooter games, there has not been much convincing evidence that it has a detrimental effect on people, let alone any effect. Although several fMRI studies have suggested that there are alterations to the default networks in people who are addicted to video games, another study failed to find any effect of violent video games on resting-state patterns in nonaddicted players.[14] But the field is still young. I suspect that in the future we will find that video games, especially first-person, will have demonstrable effects on the brain, especially during adolescence.

For now, books remain the most effective way to change the narrative trajectories in our brains. The imaging results suggest that it is the immersive nature of the reading process that taps into the brain's imagination system. As we have seen with the replay of memories, the brain repurposes its sensory systems, but in the case of books, it is in service of projecting the reader into the shoes of the protagonist. A good novel puts you in the skin of another person to the point that you may actually feel like them. The imaging results show that such a simulation appears to leave a trace for, at least, a few days. But with the right material, I believe these changes may be more persistent.

These findings raise the intriguing possibility that the stories you consume, especially the ones you read, are food for the mind. This dietary analogy goes further: You are what you eat. The stories you consume become part of you, and the repeated stimulation of the sensory centers begins to form the narrative equivalent of muscle memory. Your brain gets accustomed to these narrative archetypes. It may not matter that you know, at some level, that they're fictions. They still reinforce the templates in your brain that you call up for interpreting the events of your life. The flip side is that you have control over what stories to consume. Stories of heroes will reinforce the feeling that you, too, are on a hero's journey. But as we'll see in the next chapter, a steady diet of stories featuring conspiracies and shadowy forces can nudge your personal narrative in the other direction, viewing the world through a lens of suspicion and paranoia.

CHAPTER 15

Hijacked Narratives

I N PART II, WE LEARNED ABOUT THEORY OF MIND (ToM), an ability that evolved to let us simulate and consider what other people are thinking. ToM allows us to solve coordination problems like the stag hunt and generally makes living together easier. At the same time, by making it easy to adopt others' points of view, ToM provides a back door into our brains, and if we are not vigilant, all sorts of bogus facts can get intermixed with ongoing streams of information, hijacking our narratives. If a person continues with a diet of junk information, they risk unwittingly altering their sense of self. At scale, this can become a real problem. If enough people believe a delusion, then it ceases to be a delusion.

We perceive and understand everything through the narratives we absorb throughout our lives. Surround yourself with people who tell fantastical stories of alien abductions and CIA cover-ups, and you will begin to see the world that way. In this chapter, I examine how certain types of stories can sneak their way into our brains and hijack our personal narratives without our even realizing it. I'm talking, of course, about conspiracy theories. Whatever your views on conspiracy theories, there exists a large body of scientific knowledge about how they form and why so many people come to believe them. And perhaps counterintuitively, conspiracy theories can offer even the most skeptical among us an important lesson in how to rewrite narratives, by showing us how easy it is to turn a story on its head and see the world in a different way. It is an effective technique that can be used to reinterpret and rewrite any narrative, including one's own.

But first, we must descend the rabbit hole.

On May 4, 2020, roughly two months after Covid-19 had begun shutting down the US economy and sending people into lockdown for fear of contracting the disease, a twenty-six-minute teaser video was released. *Plandemic*, as it is called, offers up a terrifying story of how the scientific and medical establishment, led by Dr. Anthony Fauci, planned the coronavirus outbreak so as to leverage vaccine treatments for which they were holding patents. It immediately went viral. The star of the video, Dr. Judy Mikovits, was already well known to virus researchers as a vocal anti-vaxxer. Her appearance in this controversial video surprised nobody familiar with her story. The entire saga is a case study in the construction of false narratives and conspiracy theories.

The pitch for *Plandemic* goes something like this: Imagine that a cabal of physicians and researchers are hiding the truth about coronavirus. In this scenario, Covid did not arise naturally, but it was concocted by researchers in the US and Chinese governments who stood to profit from the forced vaccination of the American public.[1]

You may ask, what kind of person would make up such a story? The producer, Mikki Willis, describes himself as a father of two boys living in Ojai, California, and is the founder of a small production company called Elevate. According to his IMDb page, "Elevate is one of the most prolific creators of socially conscious media. As a filmmaker Mikki has been granted rare access into the minds and hearts of world re-known philosophers, scientists, physicists, doctors, authors, shamans, and human development experts."[2] It does not mention *Plandemic*, which was banned by all of the major social media platforms two days after its release. In addition to producing the video, Willis plays the interviewer. Throughout the conversation, Willis, a former model with sea-blue eyes, drips sympathy as Mikovits describes the mistreatment she endured by the scientific establishment. Mikovits is happy to play the martyr, telling Willis how she was jailed without being charged, allegedly for removing laboratory notebooks and proprietary information from a research center in Reno, Nevada.[3] But the implication is that she was arrested for her beliefs about vaccines.

Plandemic is undeniably slick. Despite costing Willis only $2,000, the production values are high. The professional look gives it an air of authenticity, not quite *60 Minutes* level, but better than a lot of what you can find on Netflix. Although Willis promoted it to be shocking and conspiratorial, he later appeared surprised at how quickly the video went viral. Facebook and YouTube quickly removed the video for "suggesting that wearing a mask can make you sick."[4] That didn't stop its spread. If anything, it accelerated matters by playing into the conspiracy theory. It is impossible to know Willis's and Mikovits's motives, whether they intended to create a video that would potentially endanger thousands of lives. Yet intentionally or not, *Plandemic* checked all the boxes for sucking people into a web of disinformation.

Countless similar narratives have been promulgated throughout history, long predating the internet. What made *Plandemic* so successful? The answer lies in the science of conspiracy theories. What follows is not meant to be an instruction manual for creating them. Think of it instead as a form of inoculation against such theories, as well as a primer on how to use similar strategies to change one's own narrative.

PLANDEMIC HAS ROOTS IN A PARTICULARLY POTENT IDEOL-ogy: the anti-vaxxer movement. Anti-vaxxers became particularly vocal in the early 2000s, when the movement gained momentum through high-profile celebrity activists, such as actor Jim Carrey and his ex-girlfriend Jenny McCarthy. After California governor Jerry Brown signed a bill in 2015 that removed the state's vacci-nation exemptions, Carrey tweeted, "California Gov says yes to poisoning more children with mercury and aluminum in manda-tory vaccines. This corporate fascist must be stopped." He went on to criticize the CDC about the use of mercury and thimerosal in the vaccines, "The CDC can't solve a problem they helped start. They are corrupt."[5] McCarthy, meanwhile, has stated that the MMR (measles, mumps, rubella) vaccine caused her son's autism. And long before his presidency, Donald Trump was tweeting rants against vaccines, calling them monster shots.[6]

One of the movement's pet causes is the claim that vaccines cause autism, and the modern rise of anti-vaxxers is correlated with the rise in autism. It is a fact that autism spectrum diagnoses have increased dramatically. According to the CDC, in 2000, the prevalence rate for autism spectrum disorder (ASD) was 1 in 150 children. By 2016, it was 1 in 54.[7] Parents of these kids began looking for an explanation and found it in a small-sample case history of children with ASD in which the onset of their symp-toms occurred after receiving the MMR vaccine. The study was eventually retracted because it only followed a sample of twelve kids, and, as every scientist knows, correlation does not mean causation. Yet the study caused a dramatic rise in parents not vac-cinating their children and also those selecting which vaccines to give—the slow-vaxxers.

The anti-vaccination movement didn't begin with the MMR vaccine. Anti-vaxxers have been around since the very first vaccine was discovered, even as it effectively prevented one of the most feared and deadly diseases at the time—smallpox. In 1796, Edward Jenner tested his vaccine on the eight-year-old son of his gardener and proved that it worked by later exposing him to the smallpox virus. It was a stunning achievement, and within the next ten years, vaccinations had become widespread in England. However, some people were offended by the perceived uncleanliness of having pus derived from an animal rubbed into cuts in their skin, making it, they believed, un-Christian.[8] And in a theme that echoes in the present day, others objected to violations of their personal liberty when the British government ordered mandatory vaccinations in 1840. In the face of mounting protests, the British Vaccination Act was amended in 1898 to allow for conscientious objectors, which was the first time the term had been used.[9] Similar movements were afoot in the United States. The Anti-Vaccination League of New York City was founded in 1885. In 1905, the US Supreme Court ruled that Massachusetts could enact compulsory vaccination laws in the interest of public health. In the 1970s, rates of vaccination had decreased in Britain, including the standard diphtheria-tetanus-pertussis jab all kids were supposed to receive. In part, this was due to confusing reports about neurological disorders linked to DTP.[10] And then the MMR controversy hit in 1998, not coincidentally also originating with a British physician, Andrew Wakefield.

Viewed within its historical context, the modern anti-vaxxer movement rehashes the same fears and concerns about individual freedom that have been the core of its ideology since the days of Edward Jenner. There is often no reasoning people out of such beliefs, especially as they are often underpinned by spiritual and philosophical justifications. Anti-vaxxers tend to be highly educated, and a potent mix of views about the environment, healing, new-age spiritualism, and individual liberty underlies the ideology. What began as an isolated opinion about a single vaccine starts to sound like religion, or a mass delusion.

These factors—environment, healing, and liberty—tap into fundamental fears and thus effectively act as back doors into the human mind. Indeed, a 2010 analysis of the content of anti-vaccination websites revealed common themes.[11] One hundred percent of the sites raised concerns about safety, particularly poisons and unexplained illnesses like autism. Eighty-eight percent favored alternative medicine while pushing the idea that vaccines eroded immunity. Seventy-five percent brought up civil liberties, and 75 percent raised the specter of conspiracy theories. These included Big Pharma making a profit, vaccine promoters causing the disease they purport to protect, and cover-ups. Fifty percent pointed to rebel doctors who bucked the medical establishment. *Plandemic* ticked all these boxes, so it is not surprising that it gained traction.

Another reason for the explosion in anti-vaxx sentiment is who's promoting it. As is the case with many narratives, most of us don't have access to the facts about vaccines and have to rely on the narrator. Which brings us back to the epistemological question of how we know what we know. In lieu of proof of fact, it is natural to substitute the perceived truthfulness of the narrators themselves. A narrator's authority will be influenced by their reputation and the skill with which they tell their story. We have already seen how readily our brains incorporate other people's narratives and how an expert's opinion can result in the offloading of hard thinking. As a scientist, and an honest human being, I would like to avoid ingesting false or misrepresented narratives, but sometimes it can be hard to tell when it is happening.

THERE ARE CLUES TO IDENTIFYING FALSE NARRATIVES, BUT they can be subtle, and one has to be constantly vigilant when looking for them. Karen Douglas, a psychologist at the University of Kent, has provided a useful framework for analyzing conspiracy theories in particular, and which we can apply to false narratives more generally. She identifies three motives for why people are hijacked by conspiracy theories: epistemic, existential, and social.[12] The epistemic motive for conspiracy theories builds on the process of belief. A belief is an attitude that a person holds regarding something they think is true. It can be based on facts or it might exist without proof, like a belief in the existence of God. In Chapter 2 we saw that the rationale for a belief is called a justification. The question of justification is philosophically and practically important because it gets at why a person holds a particular belief. You might believe something because you saw it with your own eyes, you logically deduced it, or someone else told you about it.

Conspiracy theories are appealing because they create an explanation for events that are otherwise unexplainable. This is the back door I alluded to. Many false narratives provide explanations based on hidden actors. These narratives can be seductive because the idea that there are cabals of people working out of public view to bring about a particular event is nearly impossible to disprove. The political scientist Russell Hardin calls this form of justification a "crippled epistemology," meaning that a person doesn't know much, and what they do know is wrong.[13] Cass Sunstein and Adrian Vermeule, also political scientists, expanded on this idea by pointing out that crippled epistemologies often underlie extremist behavior, sometimes to the point of terrorism.[14] For most beliefs, people lack direct knowledge, so they must rely on other people as their sources. Extremist views are not necessarily irrational but stem from a paucity of information and reliance on a limited social network.

When bad things happen—like 9/11 or the Covid pandemic—desire for information is high, but accurate reporting is in short supply. People naturally turn to their social network for intelligence, which leads to rumor and speculation, providing the perfect growth medium for conspiracy theories. Sunstein and Vermeule have argued that the lack of information merely serves as fuel for conspiracy theories. Other factors are required to ignite the fire and maintain it. High-arousal emotions like fear and anger fan the flames. Rumors and theories feed on these emotions. A conspiracy theory provides the soothing appearance of a rational justification for what is otherwise an unexplainable feeling, a process called emotional snowballing.[15] Conspiracy theories also provide a bulwark against naked existential fear. Because conspiracy theories feed on rumors rather than facts, they promote group polarization into believers and nonbelievers. Sometimes, people don't necessarily believe a conspiracy theory but will go along with it because their peers do and they don't want to be cast out. As more people buy into a conspiracy theory, either explicitly or tacitly, the conspiracy cascades, giving rise to the perception that if everyone else believes it, it must be true. As we saw in Chapter 10, the law of large numbers says that other people are probably correct, and our brains are primed to believe it, providing another back door to being hijacked.

Although conspiracy theories initially arise because of a lack of information, particularly in response to a frightening event, they feed on themselves through social mechanisms. In addition to validating the self through group identity, conspiracy theories almost always adopt a specific form of narrative—one that is already deeply embedded in everyone's brain. A conspiracy theory is a form of journey in which external forces impel a hero to right a wrong. But here is why they are so pernicious: Because a conspiracy theory assigns blame for the tragic event to someone outside of the in-group, it relieves the believer of any culpability. It is an act of collective narcissism that places believers above nonbelievers, and the narcissism is amplified by the conviction that believers are not truly appreciated.[16] A conspiracy theory is the perfect underdog story, pitting scrappy insiders (the ones who know the "true" story) against the disbelieving outsider majority. Inevitably, conspiracy theories erode trust in established institutions while creating new, alternative ones.

PLANDEMIC TAPPED INTO ALL THREE OF THESE THEMES. IT built on the extant fear of vaccination as a conspiratorial explanation for illnesses like autism, whose cause is presently unknown, and melded it with the novel fear of coronavirus. Most people felt helpless in the face of the virus, but rather than confront these fears rationally with science and evidence, *Plandemic* displaced personal responsibility with blame, pointing the finger at nefarious government actors who wanted to profit from a future vaccine. Anthony Fauci made for a perfect villain: he was an erudite government scientist who was named on various patents (unrelated to coronavirus), and he had previously tussled with Judy Mikovits— the hero of the film.

Plandemic opens by introducing Mikovits as a martyr for the truth. Willis, the producer and narrator, sets the stage with this dire warning:

> At the height of her career, Dr. Mikovits published a blockbuster article in the journal, Science. The controversial article sent shockwaves through the scientific community as it revealed that the common use of animal and fetal tissues were unleashing devastating plagues of chronic diseases. For exposing their deadly secrets, the minions of big pharma waged war on Dr. Mikovits, destroying her good name, career, and personal life. Now, as the fate of nations hangs in the balance, Dr. Mikovits is naming names of those behind the plague of corruption that places all human life in danger.

Putting aside the bombastic framing for now, in the interest of fact checking, it is true that Mikovits published a controversial article in *Science*, but it was on an unrelated virus, called xenotropic murine leukemia virus-related virus (XMRV), and the disease she linked it to was chronic fatigue syndrome—already controversial because physicians and scientists couldn't agree on whether CFS constituted a physical illness or whether it was a mental disorder akin to the hysteria of Freud's era. Nobody could reproduce Mikovits's results and, when pressed further, the authors acknowledged problems with the technique they had used. The editors of *Science* retracted the article two years later.[17]

The retraction of a scientific article is a traumatic event, especially for the senior author, who is deemed ultimately responsible for the article's contents. To be fair, mistakes happen. A retraction can be a simple mea culpa, after which everyone moves on. In some cases, though, the authors dig in their heels. That is what Mikovits did, which prompted *Science*, as Willis states, to publish a hatchet job on her work. In an unusual eight-page exposé, the journal editors laid out all the gory details of her controversial work, which were of interest primarily to those in the field. Mikovits was quoted as responding, "I don't care if nobody else in the world wants to work on it. Fine, leave us alone!" Another researcher said presciently, "I began comparing Judy Mikovits to Joan of Arc. The scientists will burn her at the stake, but her faithful following will have her canonized."[18]

Martyr narratives have deep roots. In Greek, *martyr* refers to bearing witness, that is, someone who speaks of the knowledge they have. In Christianity, this means the word of God, with martyrdom a direct path to sainthood in the Catholic Church. Originally, though, there was nothing religious about bearing witness. In Aristotle's view, a martyr referred to a virtuous man who always spoke the truth. Of course, this won't make you many friends. Those who are outspoken, whatever the cause, are bound to draw the ire of their peers. Those who bear witness are often shunned from society. In the not so distant past, they were often killed.

Martyrs elicit opposing reactions. On one hand, their views might conflict with the norm, making them socially repellant. As we saw in Chapter 11, societies depend on the maintenance of social norms, and to do so a few righteous people must step up and punish violators. For this reason, martyrs will be shunned by society. At the same time, martyrs garner (often posthumous) respect for their courage to speak what they believe to be the truth and their willingness to see it to the bitter end. They are the ultimate hero. Consider the archetypes: Socrates, Jesus Christ, John the Baptist, Joan of Arc, Abraham Lincoln, Mahatma Gandhi, Malcolm X, Che Guevara, Martin Luther King Jr., Nelson Mandela.

There is no single kind of person who becomes a martyr, as psychologists discovered when they developed a "self-sacrifice scale." A willingness for self-sacrifice, it turns out, is not correlated with any particular personality type.[19] This suggests that people come to martyrdom through life experience rather than genetics. Scott Atran, the cultural anthropologist, believed that humiliation played a powerful role in birthing martyrs and terrorists. This certainly fits with the *Plandemic* narrative. Mikovits was publicly humiliated. Under such circumstances, it would be more surprising if she didn't hold a grudge.

All of this is to say: beware the martyr narrative. Even as the martyr speaks truth to power, they likely harbor a deep-seated grudge based on perceived wrongs, which can warp their judgment. The martyrs I have mentioned in this chapter are uncommon in that they reached a level of notoriety that the everyday martyr doesn't. But martyrdom is one of the six narrative basis functions I described in Chapter 8 and well known to everyone. Everyone has the potential, from time to time, to act the martyr in their personal dramas. It is an attempt to put a positive spin on the riches-to-rags story, converting it to a man-in-a-hole story.

Martyrdom, then, is more common than we like to acknowledge. It is really a matter of degree.

Throughout this book, I have discussed how narratives are constructions that convey a sequence of events, and how there are several potential narratives for the same sequence. Without any corroborating information, it can be difficult to tell whether a particular narrative is true or false. Conspiracy theories, especially those featuring a martyr figure, offer a potent lesson in how to deconstruct a narrative. The same tools can be applied to all narratives that are potentially suspect. Here is a brief checklist:

- Credibility of the narrator. What is their authority or expertise? Do they have an agenda?
- Does the narrative push a grievance? Some grievances are legitimate, especially those based on objectively verifiable harms to a group of people (e.g., enslaved people, indigenous peoples). Others, though, are based on personal grievance. Note any past slights and humiliations.
- Source the epistemology. Ask how the narrator knows what they know.
- Finally, beware the martyr.

These questions may make you wonder whether a person pushing a false narrative knows that it isn't true. Whereas we may forgive people with crippled epistemologies for passing along false narratives out of ignorance, we tend to be less magnanimous toward those who knowingly promote falsehoods. In the last chapter, I argued that you are what you eat. Conspiracy theorists tend to be an extreme example, transforming consumers into promoters. While dabbling in the occasional conspiracy theory is probably harmless, if you spend hours each day searching the internet for like-minded people, then you risk developing a crippling epistemology—or what the media likes to call an echo chamber. Under these circumstances, there is no ground truth. There is no external frame of reference. Confirmation bias to the extreme.

There is a silver lining, however.

When you realize the power of some false narratives to hijack your identity, you can't help but wonder whether you could be as easily compelled by narratives of a different genre. Are there alternative narratives that feed into our journey motifs but in a more productive way? Something that could make you a different, better person?

I believe there are. In the final few chapters, I will suggest strategies for crafting new narratives that change both who you think you are and who you can become.

CHAPTER 16

No Regrets

I'VE TALKED A LOT ABOUT HOW CONSUMING DIFFERENT types of stories creates templates for understanding our sense of self. Hero stories promote a sense of self-as-hero. Conspiracy theories cement an underdog mindset and pit Us against Them. Yet what about alternative futures? In this chapter, we will see how past-selves ripple into future-selves. This will be a targeted journey of sorts, focusing on what I call branch points in our narratives. These are the forks in the road of life, the points in time where we make choices to which we attach outsize importance in who we think we are. These include the commonly agreed-upon Big Decisions like college and career choices, relationships, and where to live. They also include choices that at the time seem inconsequential but in hindsight become significant. For instance, a last-minute decision to go to a friend of a friend's birthday party where you meet a stranger—who later becomes your spouse. The psychologists Daniel Kahneman and Amos Tversky have called these "fault lines" and found that people tended to focus on these same types of events.[1]

What unifies these branch points is the looming possibility of regret. Interestingly, regret ripples both backward and forward in time. Everyone has things they wish they could change about the past. Maybe it's the person you treated poorly in high school, or maybe it is the thing never done because you were too busy or too scared. Regrets can result from acts of commission as well as omission. But regret gets projected into the future, too. We each have unique collections of regrets, and if we learn anything from them, it is to do our best to avoid them in the future. Here, the predictive brain kicks into action. We make decisions with an eye to our future-selves looking backward in time, doing our best to avoid future regret.

Regret is a perverse form of time shifting. It may seem so convoluted that surely only humans inflict this form of mental torture upon themselves. Not so. In fact, regret is a powerful form of learning that evolution has discovered and imbued all animals with. Regret is deeply ingrained in our biology. But that doesn't mean it can't be harnessed to reshape our ongoing and future narratives.

As powerful prediction engines, our brains do a fair amount of time shifting, scooting forward and backward from the present to the future and occasionally dipping into the past for guidance. The system operates even when we're not thinking about it. Without it, you wouldn't be able to cross a street without getting hit by a car. Prediction systems come to the front of mind when we have to consciously make choices. The weightier the decisions, the more obvious it becomes that we peer into the future and try to imagine what it will feel like under different potential outcomes. It's not too different from what a fiction writer does when crafting a plot outline. And there is always the nagging concern of *what might I lose?* Nobody wants to make a choice that leads to grief down the road.

It is important to distinguish between disappointment and regret.[2] Disappointment occurs when things don't go one's way, but the outcome has nothing to do with the person's choices. Disappointment is the feeling you get after being turned down for a job or a college. Assuming you've done your best, the feeling arises from wishing that the state of the world were different. Psychologists call this *counterfactual thinking*. When we are disappointed, we engage the counterfactuals of alternative states of the world, as in, *I am disappointed that I live in a world of global warming.* I can imagine alternatives, but none of them are within my control.

A different form of counterfactual thinking occurs when we think about our choices. Instead of not getting the job, now imagine you have two job offers. One offers more money than the other, but it requires moving to a new city. Let's say you go for the higher pay, but five years later find that you hate the job and the city you moved to, and you wish you had taken the other one. That is regret. It is the consideration of alternative states of the world had you made a different choice.

In recent years, both neuroscientists and economists have turned their collective attention toward the study of regret. This may seem like an odd pairing, but regret is a psychological phenomenon that can be written down as an equation. Any human decision that can be turned into math will appeal to economists. Similarly, a cognitive process that can be reduced to an equation can be tested by looking for its correlates in the brain.

Without belaboring the math, the regret equation goes like this. Imagine that you are at a branch point, and you have to select one of two options. Each choice might eventually lead to some benefit in the future, but the outcomes are by no means certain. Standard economic theory says you should pick the option with the highest expected utility, meaning its magnitude multiplied by its probability of occurrence. But thinking about your future-self, you know that if things don't turn out the way you hoped, you would agonize over the other pathway, what might have been. The equation quantifies regret as the difference between *what is* and *what might have been*. And if it turns out that you made the right choice? Then *what is* is better than *what might have been*, and that is called *rejoice*.

In addition to experiencing counterfactual outcomes as relative amounts of regret and rejoice, the theory suggests that people make decisions with an eye to minimizing future regret.

Several brain imaging studies have been undertaken to reveal what, if any, brain circuits might be involved in the experience of regret as well as its anticipation. In 2004, a team of neuroscientists from the National Center for Scientific Research in Bron, France, tested the hypothesis that a portion of the frontal cortex, called the orbitofrontal cortex, was critically involved in the experience of regret.[3] They presented participants with a series of roulette wheel choices. Instead of picking a number, the subjects were presented two wheels and told to choose which wheel to play. In the control condition, the researchers spun both wheels but only revealed the outcome of the one the subject had chosen. In the Regret-Rejoice condition, the same procedure was followed, except the outcome of the unchosen wheel was also shown. The participants rated how they felt at the end of each trial. In the control condition, subjects were generally happy when they won even a small amount of money. However, this turned to unhappiness in the experimental condition if they saw they could have won more money. That aligns with regret theory: a better counterfactual turns something positive into a negative. The researchers then tested a small group of patients who had suffered a stroke to the orbitofrontal cortex. Remarkably, these patients showed no evidence of regret. Their ratings were the same regardless of whether or not they saw what might have been. In a separate study of normal subjects, the researchers used fMRI and the same gambling game and found that the orbitofrontal cortex was, indeed, active during the experience of regret.[4]

Read Montague, a neuroscientist at Baylor College of Medicine in Houston, used fMRI to obtain similar results. Montague's team had people play an investment game in the scanner by showing them snippets of historical stock market prices.[5] The subjects were then allowed to place bets, after which they were shown what the market did and whether they won or lost money. Montague's team defined what they called "fictive error" as the difference, in hindsight, between the best bet and the actual bet. With this mathematical equation, they showed that the caudate nucleus, which is at the heart of the reward system in the brain and connected to the orbitofrontal cortex, contained signals for both the experienced reward and the fictive one in the form of counterfactuals.

In addition, there is growing evidence that other animals experience something akin to regret.[6] David Redish, a neuroscientist at the University of Minnesota, developed something he called "restaurant row for rats." This was a circular track with four radiating spokes, and the end of each spoke contained different flavored foods: banana, cherry, chocolate, and unflavored. When a rat entered a spoke, a series of tones was played, decreasing in pitch every second. If the rat stayed until the end of the countdown, it received the flavor in that spoke. But if the rat left the zone, that flavor would become unavailable to it. An impatient rat, choosing not to wait for its favorite flavor, might discover that the next choice required it to wait even longer for a less desirable outcome. In those circumstances, the rat often looked back to the forgone option. During those wistful glances, Redish discovered that neurons in the rat's orbitofrontal cortex and caudate were highly active, just as in the brains of humans experiencing regret. These results suggest that rats, like humans, simulate what might have been.[7]

WHAT ARE WE TO MAKE OF THESE RESULTS? THE BRAIN data in both humans and rats suggest that there has been strong pressure in the evolution of mammalian brains to consider counterfactual outcomes. It is easy to see why this would be advantageous. Such thinking allows an animal to learn not only from their own decisions but also from an array of possibilities that might have happened. Your brain comes hardwired to think about the multiverse of yous. So rather than thinking of regret as some quirk of the human condition, it is more accurate to think of it as a feature of a long evolutionary history of decision-making. Every animal before you in the tree of life has experienced regret. Those who didn't died long ago.

Evolution only selects for processes that enhance survival, but regret, by its very nature is a backward-looking emotion. Yet pining over the past better prepares us for the future. We, and other animals, learn from our mistakes so as not to repeat them. Counterfactual learning is particularly effective when some choices might have fatal consequences. In such circumstances, it is far better to simulate what might have happened than to actually experience it!

I am reminded again of the bike-riding incident from my adolescence. As I described in Chapter 6, the memory is dissociated from my body, and in it I see myself from a third-person perspective. From this vantage point, I also see what might have happened had I been going a little faster or had the driver not swerved at the last moment. With the same vivid realism, I can see myself splattered across the truck's grille. It is a powerful counterfactual. Still, some forty years later, this fictive signal embedded in my brain forces me to pay attention to trucks every time I go bike riding, even if they are on the opposite side of the road. Each time I get on the bike, I do my best to avoid that branch point ever again. I didn't have to experience a horrific accident to learn from it. That is powerful learning indeed.

Although it seems plausible that evolution has guided animals to experience regret for its survival advantage, it has been difficult to find concrete evidence for this view. Remarkably, new evidence from the field of artificial intelligence supports the adaptive value for individual regret. One popular way that computer scientists test their algorithms is to hold tournaments. These competitions provide a hyperaccelerated environment for evolution, selecting winners and killing off losers at lightning speed. One such contest is the Annual Computer Poker Competition. Poker is a game with imperfect information, meaning the players never have all the information they would like because the other players hold some cards privately. These types of games are often harder for computers to solve than chess, a game of perfect information where every move is played in the open.

Shortly after the Computer Poker Competition began in 2006, a team of computer scientists at the University of Alberta, Canada, introduced a new algorithm based on "counterfactual regret minimization," or CFR.[8] Instead of trying to simulate all possible decision trees in a game, like chess algorithms do, CFR has the computer play against itself internally. In these simulated games, the computer searches for strategies that minimize regret, defined as the difference between what happened and what might have happened. The CFR algorithm doesn't aim to solve every possible outcome. Instead, it samples the space of strategies and the average regret associated with each. It is highly efficient. It doesn't need to be perfect, and CFR doesn't take much memory because the results are stored in a compressed format in the form of averages. Versions of CFR have consistently won the tournament,

and its new-and-improved cousin, CFR+, has even solved Texas
Hold'em, which requires more frequent use of strategic play like
bluffing and going "all-in."[9]

REGRET THEORY IS A MATHEMATICAL ABSTRACTION FOR THE
counterfactuals that we experience and anticipate. Because it is
a simple equation, expressing the difference between what happened and what might have happened, it does not distinguish
between regrets of commission and regrets of omission: things
we've done versus things we wish we'd done. Psychology, though,
reveals that there are differences between the two.

The psychologists Thomas Gilovich and Victoria Medvec proposed that actions—that is, acts of commission—are more likely
to induce regret in the short term, but this decreases quickly with
time. On the other hand, inaction—or an act of omission—may
lead to worsening regret over the long run.[10] When people are
asked about their biggest regrets in life, Gilovich and Medvec
noted that they often focus on the things that they failed to do, for
instance: *I wish I took that trip to Europe*; or *I regret I never went on a
date with so-and-so*; or *I wish I told my father I loved him before he died.*
Almost every study that has examined this question has come to
similar conclusions. The most common regrets of failure to act are
missed educational opportunity; failure to seize the moment; not
spending enough time with friends and relatives; missed romantic
opportunity; not pursuing interest X.

Gilovich and Medvec identified several factors that contribute
to regrets of omission. First, the passage of time tends to increase
the regret because each time a person thinks about the missed opportunity, they have a tendency to become more sure that their
counterfactuals are better than what actually happened. This is
illusory, though, because there is no way to know what might have
happened had a particular action been taken. Second, as one's

confidence in the alternative reality grows, so does the inexplicableness of not acting. The more inexplicable the failure to act, the more likely that shame gets layered on top of the regret. Finally, whereas the consequences of a regrettable action are circumscribed because the outcome is known, the consequences of not acting are potentially infinite. The number of possible alternative realities is limited only by one's imagination.

Regardless of whether they are acts of omission or commission, regrets have the power to rewrite a person's narrative. They can provide meaning or a sense of purpose to events that, in fact, you might not have had any control over. Although I frequently think of the bicycling incident, if I am truthful with myself, I never really had control over the situation. I am grateful that I wasn't killed, but that had more to do with the truck driver than me. Even so, it stands as a branch point in my life, and I ascribe meaning to it. Interestingly, the psychologist Keith Markman used the very scenario of almost getting hit by a truck to study the ways in which people form counterfactuals.[11] In this case, the counterfactual is worse than what actually happened, so it is called a downward counterfactual. Even so, Markman found that people tended to view it in one of two ways. You could say, "I was nearly killed by that truck," or you could say, "I was lucky to not have been hit by that truck." The first, which he called a downward reflection, induces a negative mood state, while the second, which he called a downward evaluation, is associated with a positive mood.

It is important to remember that counterfactuals serve two purposes. Even if they are made-up, some counterfactuals explain the past. This is where narrative explanations come in. If I was religious, I might say that God was looking over me that day. At the same time, counterfactuals help us prepare for the future.[12] Here, too, narrative construction comes into play in the form of imagined regrets and rejoices of our future-selves. Markman's research suggests that some ways of thinking about counterfactuals may be more effective at changing future behavior. Downward evaluative assessments are not particularly useful because they merely accept what happened. A downward reflection, as I experienced, is more likely to promote preventive behavior in the future.

Rejoice, the opposite of regret, also figures into future decisions. Like regret, these counterfactuals can be either reflected upon or evaluated. Consider these statements from Markman: "I almost got an A" or "I got a B and failed to get an A." The second statement, an upward evaluation, tends to be more activating and impels people to try harder in the future.

BECAUSE REGRET IS SUCH A POTENT FORCE, IT SHOULD BE no surprise that many movies hinge on a character's regrets. Director Richard Linklater placed regret as the central theme of his masterful trilogy *Before Sunrise* (1995), *Before Sunset* (2004), and *Before Midnight* (2013). In the first film, Jesse (Ethan Hawke) meets Céline (Julie Delpy) on a train from Budapest. Jesse convinces Céline to get off at Vienna and spend the night roaming the city with him. In the purest portrayal of future regret, he tells her that if she doesn't go with him, in ten or twenty years she might look back on the moment and wonder how her life would have been different. The movie ends with plans to reconnect at the train station in six months. In the second film, which takes place nine years later, we learn that they never met up again. Instead, Jesse has written a best-selling novel based on his encounter with Céline. While on a book tour, he meets her again in Paris. For the rest of the afternoon, they walk around Paris, expressing their regrets about how their lives might have been different had they actually met up again as they had promised. In the final film, nine years again after the second, Jesse and Céline are married. Again, they are at a branch point in life, this time questioning their love for each other. Jesse uses his trick of time shifting into the future and looking back on the present moment, concluding that they will think of this night as one of the best of their lives.

In *Casablanca*, Rick (Humphrey Bogart) runs a shady nightclub, all the while pining away for Ilsa (Ingrid Bergman), the woman he fell in love with in Paris the year before. Ilsa had left Rick suddenly and without explanation, which caused Rick to become the gruff, cynical club owner we meet in the movie. Only later does he find out that Ilsa had been married to Victor Laszlo and that she thought her husband had died in a Nazi concentration camp. When Ilsa learned that Laszlo had escaped, she fled Paris to meet him without telling Rick. When Ilsa and Laszlo walk into Rick's club a year later, Rick has to decide whether to help them escape to America. Ilsa and Rick are still in love, and Laszlo knows it. *Casablanca* is one of the most popular films of all time for good reason. Even though the dialogue can seem cheesy, the core theme of a doomed romance still resonates. Even though Rick does the right thing and lets Ilsa and Laszlo go, the overriding feeling we are left with is *what if?* What if Laszlo had really died? Would Rick and Ilsa still be together?

These films serve as master classes in harnessing anticipated future regret to make decisions in the present. They offer powerful lessons for how regret can be used productively to shape future narratives. Rather than dwelling on the past about things that might have been, we can time shift into the future and use simulated regret to make decisions right now.

The mantra "No Regrets" turns out to be a very effective, maybe even the best, life strategy. It works for computers playing poker. It works for romantic movie characters, and it will work for you. Keep in mind that when we place stock in what are, essentially, fictional narratives about the future, some versions are better than others. In the next chapter, we will try to separate the good ones from the bad and set forth on a journey to write the narratives unfolding into the future. Everyone wants to live a life full of meaning, but that is a direct result of the narrative a person constructs. It is an active process, requiring constant attention to the interpretation of events as they occur and placing them in the context of who you have been and who you want to be.

CHAPTER 17

The Good Life

OVER THE COURSE OF THIS BOOK WE'VE EXCAVATED THE self delusion and taken a tour of your multiple selves. We've learned your own personal narrative is far from accurate, shaped by memories compressed along canonical experiences called basis functions, which in turn color anything that you experience. At the same time, your Bayesian brain creates the most likely interpretation of any event, filling in any holes with fictions. Meanwhile, your perceptions are colored by what other people think, and ideas that you believe original to you most likely originated elsewhere. All of which returns us to a version of our original question: Which is the true you? The version you think of yourself, the version others think of you, or the version you tell other people?

The answer, of course, is all of them. I like to think of the task of narrative construction as analogous to that of a film director. We collect a slew of clips from life, which we constantly edit to form cohesive narratives. The director's job is to deliver a complete movie to the producer and distributor. The analogy breaks down, though, if you consider your life as a single film, because you probably don't want to wait until you are dead to deliver the final cut. But that is the wrong timescale to think about narrative. It is more fruitful to think of one's personal narrative as a series of films.

Think of your favorite movie sequels. Which ones endure? *The Godfather* trilogy comes to mind, but that is a multigenerational story with such a sprawling cast that it does not serve us well as a model of personal narrative. Perhaps the original *Star Wars* trilogy (parts IV, V, and VI). While grander in scale than *The Godfather*, *Star Wars* is still fundamentally the story of a single person—Luke Skywalker. We have already seen how *Star Wars* followed Campbell's monomyth formula. Other trilogies focused on an individual include *Terminator*, *Rocky*, *The Girl with the Dragon Tattoo*, *The Matrix*, *Mad Max*, *Alien*, *Indiana Jones*, and *Lord of the Rings*.

These trilogies stand out because, as the screenwriting expert Robert McKee tells us, "We tell stories about people who have something to lose—family, careers, ideals, opportunities, reputations, realistic hopes and dreams."[1] Although he was referring to the craft of screenwriting, his statement also serves as apt advice for the telling of our own stories. We're drawn to stories about people who overcome such challenges because we all have something to lose. Films offer important lessons—for both life and for neuroscience—as they reflect on and inform the human condition. They are also perfect examples of compressed narratives. Just think about how much gets rolled into two hours. We are drawn to movies because they contain elemental truths about the human condition as seen through other people's eyes. We like them because they resonate with our own ongoing narratives. Through a closer examination of what makes a good film, we can learn how to write the stories we want to live. As I've said before, you are what you eat. If you accept that who you think you are derives from the stories you tell yourself, then it follows that by telling a different version of the story, you can become someone else.

But, you may object, I am already in the middle of my narrative. I can't change history and simply start over. Actually, you can.

We are accustomed to stories having a beginning, a middle, and an end. The classic three-act storytelling structure can be traced to the nineteenth-century German novelist Gustav Freytag. By analyzing Shakespearean and ancient Greek plays, Freytag determined that they all followed a predictable pattern of rising and falling action, henceforth known as Freytag's Pyramid. It is still the dominant form of storytelling in film. I hesitate to call it a formula because it is a description, not a prescription.

Yet our personal narrative doesn't have to strictly follow our biological one. Although a human life has a natural beginning and end, that is not an interesting story in and of itself. Who wants to hear that Joe Blow was born, some good stuff and some bad stuff happened to him, and then he died? No, we want to hear about his struggles. What hurdles did he face? What did he stand to lose? And did he succeed in overcoming them, or was he defeated? And, most importantly, *what was the point?*

These are the themes of good stories because these are themes that we prioritize in our own narratives. But they still have to start somewhere. Unlike the story of Joe Blow, most good stories don't begin with the birth of the protagonist (unless the hero is named David Copperfield). Rather, they start in medias res—in the middle of things. Homer's *Odyssey*, which tells the story of the Greek king Odysseus, begins after the Trojan War. We learn about Odysseus's life as we follow the course of his epic journey home.

In our own narratives, we find it difficult to step out of our own personal sequence of events. We naturally construct a narrative that extends all the way back to our earliest memories. But suppose you could step outside of that framework. Then, you could tell a new story, one that begins today.

IF WE ARE TO BEGIN A NEW STORY, WE STILL NEED A TEMplate. We can't very well begin a story without a sense of where it is going, as such a story would read like a series of journal entries—reporting on one thing after another as they occur but without any awareness of their significance. Most of us default to a hero's journey. A fine format, if you like. But the ancient philosophers, I believe, gave us a better alternative.

The notion of a good life can be traced to thinkers like Plato, Aristotle, and Confucius. Although heroes' journeys were commonly told as entertainment, they were not necessarily held up as models for how a person should live their life. The foundation of Greek moral philosophy was anchored in the concept of *eudaimonia*, which can be roughly translated as "flourishing." It is sometimes interpreted as happiness, but the Greek notion of happiness was linked to a person's connection with society while living a virtuous life. Plato described the cardinal virtues of a good Athenian as courage, wisdom, justice, and moderation. The virtuous citizen lived a life without excess, had the courage to be true to themself, acquired wisdom through experience, and shared it with others, seeking fairness among their peers. On the other side of the world, Confucius also placed virtues at the center of his ethical framework for how people should live their lives. The Chinese *dao* corresponded roughly to the Greek *eudaimonia*.[2] Confucius recognized that *dao* derived not from individual achievements but rather from how one treated other people in everyday life. Dramatic moral dilemmas and hero journeys were not nearly as significant in ancient China as in the Western canon. Virtue ethics provides an alternative narrative to the hero's journey, which is, frankly, a high bar to clear and focused on the achievement of the individual to vanquish a foe. I also think that the classical notion of virtue is what most people desire in life.

While the hero's journey and the virtuous life provide broad templates for narratives, they do not tell us how to construct one for our lives. For that, I turn to Viktor Frankl, the Austrian psychiatrist who survived years in the Nazi concentration camps. He comes closest to offering what we're all looking for in our narratives: *meaning*.

Frankl's theory, which he called logotherapy, differed fundamentally from Sigmund Freud's psychotherapy. Whereas Freud focused on a person's past, Frankl focused "on the meanings to be fulfilled in the future."[3] The causes of existential crises, Frankl realized, are rooted in disconnections between a person's current situation and what they want to be in the future. When the present does not offer a path to the future, existential distress ensues, leaving an internal void that can be summed up by the question, *Why is life worth living?* A symptom of this, Frankl said, is the "Sunday neurosis" (known today as the "Sunday scaries"), which is the malaise that settles upon a person at the beginning of a week of unfulfilling work. The goal of logotherapy is to guide a person to find meaning in life.

How do we do that? We tell stories. To ourselves and to others. It really is that simple. Meaning derives from the narratives we create.

According to Frankl, there are three paths to finding meaning. First, a person can create something or do a deed. In terms of narrative forms, this is equivalent to the hero's journey. Second, a person can find meaning by experiencing something or encountering someone. Frankl links this path to love, which he says is the only way to understand another person to their core. The third path is through suffering. In the midst of suffering, it is difficult to find meaning, so Frankl would have people project themselves into the future. In an anticipation of regret theory, Frankl would then ask the future-self to look back and imagine what they had learned from their suffering.

Logotherapy provides an illuminating guide to writing our personal narratives. Note that I said *writing*, not *rewriting*. This is an important distinction. In the last chapter, we dealt with regret, which looks both backward and forward. I am not proposing rewriting the past. Rather, we can borrow from logotherapy and other tools to write the future. We will use the same techniques that good writers use. We won't be solving the riddle of the meaning of life. Instead, we will begin with a simpler question: *What if?*[4]

This strategy works better in the realm of fiction because it's easy to get bogged down in the dreary details of reality. Fiction frees us from the anchor of history. Remember, this is just an exercise. Let your imagination run wild.

HERE'S A THOUGHT EXPERIMENT IN DISSOCIATION: WHAT IF you could clone yourself, creating not only a perfect physical copy but also one who possessed an exact replica of your brain with all your memories?

Now you are immediately faced with your first decision in this narrative. You can stay in your current life, in which case you will be writing the story of your clone. Or your clone can take your place, and then you will be writing the story of yourself in a new life. There are, of course, meta-versions, too. Namely, what happens to the person who stays in your current life, knowing that there is another version out in the world, and vice versa. But set those aside for now, and choose your narrative conceit. Either you or the clone is going to begin a new life.

The next task is to pick a time frame over which your story will unfold. Here, we deviate slightly from the cinematic formula. Most movies are set over the course of a few days to a few weeks. There are exceptions, including the relatively rare movie that takes place in real time (*Dog Day Afternoon*), as well as films that span many years (*Forrest Gump, The Godfather, 2001: A Space Odyssey*). What we are after inhabits a kind of no-man's-land for film, anywhere from one to ten years. To do anything substantial takes at least a year but often much longer. One year may be too quick and ten seems too far into the future. Somewhere around five seems about right.

Imagine what your clone will do. They can become anything you want. Five years from now, who will they be? That is the end of the story you're aiming for, but the closing frame is probably not immediately apparent. Returning to the tools of storytelling, we must ask what the clone cares about. They have been imbued with all of your memories and values, so all you have to do is look inward. If you were the clone, what would you want to do?

I find it helpful to borrow from a system of decision-making commonly used in agriculture called Holistic Management.[5] The system is based on two principles: (1) nature functions in wholes, meaning that we must look beyond ourselves to the interconnections we have with other people and the environment—in other words, the self extended in space; and (2) you must understand your environment. Good advice for everyone. From there, one can build a set of practices to achieve specific goals. But before doing that, you need to decide what you want or, in the case of your clone, what they want.

Begin, perhaps, with quality-of-life statements. What does your clone value about their life in five years' time? Write down a few statements—this is important—in the present tense. Try to come up with at least five, with one each covering relationships, financial aspirations, physical and mental health, what makes you (or your clone) happy, and what you want to spend time on. If you are conceiving a hero's journey, then the statements should be oriented around achievements. If you are thinking in terms of virtue ethics, then they should reflect those ideals. Here are some examples—yours should reflect your own values, which your clone possesses as well:

- I have a community of family and colleagues that I can rely on for advice and help when necessary.

- My partner and I make time for each other.
- I am financially prudent but have enough resources to live comfortably.
- I have no debt.
- I take at least one vacation a year.
- I am physically able to engage in outdoor activities like hiking and biking.

You now have targets for the five-year narrative. Next, write another set of statements that list what you would need to do to bring about each of the quality-of-life statements. For example, to achieve "no debt," you would have to either increase your income or decrease your expenses. You might list: decrease spending on unnecessary items, pay down credit card balance by a fixed amount each month, work an extra shift each pay period. If this is not a high priority, then it should not be on your quality-of-life list. To take another example, to make time with your partner, you might list: we will spend one night a week together without watching television; we will make one meal a week together.

These exercises may seem obvious, but they require thought because they get to the heart of the question: What if I reoriented my life to prioritize finding meaning? We often don't take the time to think about what is important to us, so it can be difficult to actually articulate our values. If there is someone else involved (a partner, spouse, child, parent), then it helps to work on this exercise together. It is important, though, that you actually write down the statements in your own words. The act of writing (or typing) makes them concrete and creates a permanent record of your thoughts that you can later refer back to.

It might take a few passes to create a list that you are happy with. When you're done, you will have an inventory of what is important to you and your clone, as well as a simple road map of how to get there. Congratulations, you now have the bones of a new narrative going forward.

It's possible you'll look at your inventory and feel that it is impossible to achieve. You may be in an unhappy relationship, for example, which may be difficult to extricate yourself from. Rely on your clone for these situations—have them take your place and set you free. Either way, you face the fascinating moral dilemma of whether you choose to remain in your current life or not. Since nobody will know the clone is not you, your choice has no impact on other people—children, for example. From here, you could begin to write a story, although that's not necessary for the purposes of this exercise.

What is important, though, is that the quality-of-life statements and their accompanying goals provide an alternative framework for interpreting things that happen to you going forward. Remember, two processes define the narratives in our heads: compression and prediction. Compression allows the storage and replay of memories in terms of basis functions. Prediction uses the same basis functions to interpret new events and store them as deviations from the collection of narratives in your head. To make new narratives, you can nudge the system to interpret events differently using the quality-of-life statements and goals you have written down. If you have trouble doing this, ask yourself: *How would my clone see the world?*

You have, in essence, created an alter ego through dissociation. A Mr. Hyde to your Dr. Jekyll. Or vice versa. Some might be critical of the suggestion that we should work toward splitting off our personalities. The goal is not to create multiple personalities. Rather, I see the creation of aspirational alters as a useful tool for reframing perception. If a good novel can put a reader in the shoes of the protagonist, then everyone has within them the capacity to imagine what it is like to walk in the shoes of who they want to be.

Do not expect a smooth ride. There will be significant hurdles, both internal and external. Internally, you have to imagine alternative plotlines. Imagining a clone is a narrative device that can help get your imaginative juices flowing. So is starting with "What if?" As in, *What if I were to walk away from my job?* Or, *What if I were to go back to school?* Note that I did not include an obvious counterfactual, *What if I won the lottery?* Apart from being practically impossible, it is not something you have control over. *What if* statements work best when you retain agency. Only you can decide whether to keep working in a job or to do something else. Only you can decide to stay in a relationship or leave. Only you can decide to move to another city. Of course, there are costs associated with any choice, but remember the lesson from the previous chapter to keep an eye to the future for anticipated regret— both for things done and not done. Externally, these alternative narratives affect the people around you. Part of the *what if* exercise is to imagine how these choices ripple outward. Ideally, you would include the significant people in your life in this game, too. *What if I quit my job* could become *What if we quit our jobs?* Expect pushback. The exercises are difficult enough to do for yourself. Getting someone else to do them is ten times harder.

Anything worth doing is going to be difficult, so I want to close this chapter with some words of encouragement. Change is possible. The hypotheticals in the previous paragraph are hard because the thing keeping you put is fear: fear of losing what you have and fear of the unknown. But that doesn't stop you from changing your narrative. If you remember one thing, it should be this: Life is a sequence of events, but your narrative is what it means to you. You may not have control over the sequence, but you can absolutely choose how to narrate it.

CHAPTER 18

The Future of You

A T THE OUTSET OF THIS JOURNEY, I ASSERTED THAT everyone has three versions of themselves: the past-, present-, and future-yous. As time moves forward, present-you slips into the past just as it is replaced with future-you. It's like being on a train. Present-you is seated in a passenger car and you're hurtling along the track of time. From your perspective, you are fixed, and it is everything else that seems to be moving. And so it is with present-you, which is the one planted in the here and now. But as Einstein told us, it all depends on your frame of reference. To someone in a car idling at a railroad crossing, it is you who is sliding by. It is you who is changing, not the world.

Throughout this book, I have argued that the past-, present-, and future-yous really are different people. On a physical level, the molecular constituents of your body change from minute to minute and day to day. You really are not the same person you were yesterday. It is only through the construction of the narratives we tell ourselves that we maintain the useful illusion that we are, in fact, the same person.

If you accept this conceit, then it follows that narratives are simply another type of computation that our brains perform. Most of the time, this happens automatically. Under normal circumstances we don't spend time worrying about whether we are the same person we were yesterday. Throughout this book, we have encountered the cognitive processes that contribute to this automaticity. The brain is not a digital recording device, so our memories must be stored in a compressed format. First experiences and first stories lay down the templates by which subsequent experiences are compared and stored as divergences from these templates. By the same token, future expectations rely on similar brain mechanisms to anticipate deviations from the present.

Future-you is not a singular entity. It can't be, because the future is unknown. Instead, future-you is a cloud of possibilities, each representing a possible trajectory from this point forward. Compression, prediction, dissociation all come into play as we decide what stories to make our own. Because we already have a lifetime's worth of narrative basis functions in our heads, the process of replacement is necessarily slow and deliberate. You can displace old narratives with newer ones and decrease the frequency with which old ones are recalled, but it takes effort and time. In this final chapter, then, I want to discuss how.

ONE LIMITATION TO THINKING ABOUT THE FUTURE IS OUR reliance on memory. In Chapter 2, I talked about semantic memories, meaning knowledge of facts, and episodic memories, meaning knowledge of events. Episodic memories are often autobiographical, and recent experiments have shown that when we think about ourselves in the future, we often adopt the same imaginary viewpoints as when thinking about ourselves in the past. In fact, researchers have called this *episodic future thinking*.[1] Further proof comes from brain imaging experiments. The same neural structures are activated during episodic recall and future simulation, including parts of the temporal lobe and the prefrontal cortex. Even the visual cortex is activated as people try to picture what the future looks like.[2]

If future simulation relies on past memories, we can take control of this process. In the last chapter I laid out broad guidelines for constructing possible future narratives. Anyone who tries these exercises will see that they aren't easy. Even those with the most creative imaginations rely on past experiences to construct possible futures. But instead of a limitation, past memories provide a leverage point for imagining the future. The Harvard memory researcher Daniel Schacter has advocated for an approach he calls *episodic specificity induction*, or ESI. Using a semistructured interview, ESI guides people to recall specific details of a past experience. They are then asked to imagine something about the future. The induction process results in richer descriptions of whatever the person imagines.[3]

In other words, the future begins with the past. To imagine the future of you, you must look inward to your memories.

Here is one way this can work. Try to recall an instance in which you felt completely comfortable being yourself. Note that I didn't say *happy*. What I mean is a situation in which you felt at ease living in your skin, without people judging you and, hopefully, without an internal monologue second-guessing yourself. If you are like me, you might bring up memories from the formative years of young adulthood. I remember toiling away in the basement of the anatomy building during grad school. Even though I was an overworked PhD student, I was completely wrapped up in building robotic contraptions to probe the biomechanics of the body. Indeed, every memory I have of being completely absorbed in something is associated with an activity that allowed my curiosity to run free to find answers to scientific questions. Even though I also recall the frustrations, curious is who I am to the core.

The goal, though, is not simply to reminisce. Schacter's work tells us that by activating these specific memories, we're primed to imagine future trajectories in rich detail. In other words, rather than looking into the future and seeing a bank of fog, it is easier to graft a version of these memories onto a future-you. I, for instance, can see myself pursuing scientific questions, but they won't be the ones I am currently working on. My future-mes are working in completely different fields.

This is just my idiosyncratic example, and it won't work for anyone else. Your future must be constructed from your own memories. Moreover you must account for possible future regrets. It's not a linear process by any means. It works best if you begin by recalling memories that represent the best of you. Don't force the construction of the future yet. That will come later, and often when you least expect it, for example, when you're moving your body. Exercise is a great way to clear the mind of daily detritus— that's why we often get the best ideas while exercising. Just priming the brain by selectively remembering before you lace up your gym shoes is all you need to do. The rest will come as your mind empties.

The challenge is remembering the ideas. As soon as you finish exercising, the mind snaps back to the present and all the creative thoughts fall away like sweat off the skin. This used to bother me greatly. I would often have what seemed like the greatest insights when I was running, but as soon as I hit the shower, the thoughts would disappear down the drain. Now, I carry a smartphone. If an idea hits me in the middle of a run, I stop and record a voice memo. Most of the time they're silly ideas, but at least they're not lost to me.

Even though I am a big fan of exercise to stimulate creative thinking, few studies have conclusively identified any link between the two.[4] If exercise works for you, great. If not, don't sweat it. There are, of course, drugs, but I hesitate to make any blanket psycho-pharmacological recommendations because of the obvious risks. For those interested in the topic, I refer you to Michael Pollan's exploration of the subject, *How to Change Your Mind.*[5]

It is important to remember that the goal is not a work of art. It is to envision a future-you. The process goes like this: selectively recall good memories; clear the mind through exercise or other means; and record what pops into your head through a memo of some sort. Done well, the process should yield a kaleidoscope of future-yous. Because the future is unknown, they will be little more than visual fragments, a collage of vague images. That's fine. Try to capture one of the images and hold it in your mind's eye. Note what stands out. Where are you? Are you alone or with someone? What are you doing? This is another way back into the narrative brainstorming I described in the last chapter. Instead of coming up with a storyline, this exercise builds on the recall of selective memories to create possible images of a future-you. If you can hold it in mind long enough, you can begin to think about how to get there.

The distance between present-you and one of these future-yous is going to be vast. You know you have succeeded if, after holding the target future-you in mind, a pit of terror settles in your stomach. To move beyond the terror, it helps to have someone to talk to. Fear of the unknown can be paralyzing. Economists call it ambiguity aversion, meaning that humans, like all animals, have an innate aversion to situations in which there is incomplete information.

I have found only one way to break through the fear of the unknown, and that is to use the mental version of the algorithm that wins poker tournaments: counterfactual regret minimization, or CFR. Recall from Chapter 16 that this algorithm models as many future versions of the world as possible and chooses a course of action that minimizes the possibility for regret. When thinking about a future version of you, you have to consider only two possibilities: either staying on your current life trajectory or switching to one that will take you closer to your imagined future-you. To make this concrete, let's consider a simple decision-making matrix.

		STATE OF THE WORLD	
		A	B
ACTION	Stay	Better (stay)	Worse (stay)
	Switch	Worse (switch)	Better (switch)

There are two dimensions of this matrix. The first is the choice of actions, either staying on the current life trajectory or switching to the imaginary one. The second dimension is the state of the world, which refers to how the world will end up at some point in the future. There are, of course, an infinite number of possibilities. An asteroid, for example, could wipe out life as we know it, in which case this exercise is moot. But for simplicity, let's consider two versions of the world, which I have labeled A and B. In version A, you are better off staying on the current trajectory. In version B, you are better off switching. If the likelihood of both worlds occurring is the same, then CFR says that we need only look at the two worse possibilities. If you choose to switch paths, and things end up in World A, then you will be worse off than if you had stayed on your current path. Conversely, if you stay the course, and the world ends up in B, then you will be worse off than if you had switched. CFR tells us to *choose the least worst possibility*. This is how to minimize the chances of regret.

It doesn't guarantee the best outcome. It just means that your chance of regretting your choice is kept to a minimum.

It sounds backward. I'm advocating for imagining a multiverse of future-yous, with the idea of conjuring up a pathway to a better future. Yet the algorithm I just described is centered on the *worst* possible outcome, with the goal of making a decision to avoid it. Remember, regret comes in the commission and omission flavors. Only you can decide which type is worse. On average, regrets of commission tend to be immediate and short-lived. It is the regret of omission—*I wish I had done that thing*—that creeps up slowly and grows over time.

I teach this strategy to undergraduates in a class on decision-making. Many of the students are at a branch point in their lives where they are preoccupied with career choices. Some are terrified of making a wrong choice. Will they disappoint their parents? What if they aren't good enough? How will it feel to move to a different city? It is easy to get paralyzed by indecision. Regret minimization offers a remedy (if a decidedly unromantic one) for this stasis. I tell the students to imagine themselves four years in the future looking back on this moment. Choose the path that that future-self will not regret.

WE'VE COME TO THE END OF OUR LITTLE JOURNEY. I HOPE I have convinced you that there really are multiple versions of you, and that they have always been there. Once you realize the power of this statement, the ability to invent new narratives opens up before you. You are the narrative you tell. Not the one other people tell about you. Actions and deeds only matter to the extent that they are important for your narrative. As long as you remember that you are the narrator, then you control the plotline. You have to be diligent. You can't erase old narratives, but you can begin to displace them by consuming other narratives that more closely align with the one you desire. You are what you eat.

It seems fitting to close with a quote from the end of Shakespeare's play *The Tempest*. Prospero, who was marooned on an island with his daughter for twelve years and learned enough magic to conjure up a storm to bring help their way, finally promises to tell Alonso, one of his rescuers, how he did it. Alonso responds: *I long to hear the story of your life, which must take the ear strangely.*

Live your life. Tell a strange story.

Epilogue

NOW THAT WE ARE ABOUT TO PART WAYS, I WANT TO thank you for sticking with me through this journey. I hope the time you've spent with me was worth it.

Authors and readers, though, have very different temporal relationships to a book. If you skimmed, you might have spent an hour or two with this material, or if you read it cover to cover, maybe six hours. Even though this time was probably spread over several days or weeks, for you, it was but a snapshot in your life. From my perspective, though, I lived with this material for more than two years. I began work on this book in mid-2019 and spent 2020 and 2021 researching and writing. A few months in, the Covid pandemic struck. That is the reason for the cautionary note to the reader at the beginning of this book.

Future historians will look back on the Covid years just as we look back on the Spanish flu epidemic of 1918, namely, as a singular event in time. Of course, it wasn't. Covid was an event that played out over multiple years: 2019, 2020, 2021, 2022. If you were like me, you probably wondered where the time went. Your sense of time got confused. I would sometimes catch myself referring to something that happened "last year" when, in fact, it had happened two years before. It was as if time itself collapsed.

I think I know why. In Part I of this book, I cited research suggesting that our memories are encoded as punctuated sequences of events. The interstices between events, when nothing happens, collapse in the compression process. For many people, myself included, the Covid years were spent largely in front of a computer, working from home. Nothing happened there. The environment was unchanging. So it should be no surprise that time collapsed. It was like being on a train without any stops. An interstice stretched over years.

A year in, I realized that I needed to get off the train. My wife and I, by then empty-nesters, sold our house, bought a farm, and moved to the country. We turned our narratives on a dime. We became different people. How does one go from a suburban lifestyle, wondering where to get the best ramen, to raising chickens and cows and worrying about whether it's going to rain? The answer is contained in this book. I imagined a future version of myself. Had I stayed the course, I foresaw the possibility of real regret. Even though we knew next to nothing about agriculture, the prospect of doing the same old thing was even more terrifying than failing dramatically at a new one.

That is why the person who began writing this material is different from the person writing this epilogue. Does that mean this book is a multiauthor work? If you've made it this far, then you know the answer is *yes*. They happen to have the same name, and they look identical, give or take some wrinkles and gray hairs. By the time the book hits shelves, the author will be still a different person. So, if you run into someone who looks like the author in the airport, he may be slow to unpack this material. It is, after all, water under the bridge for him, and he is probably thinking about some future version of himself.

Acknowledgments

I HAVE BEEN FORTUNATE TO HAVE SO MANY WONDERFUL colleagues who have shaped my thinking about the human brain and what makes it different from any other type of brain on this planet. For over thirty years, Read Montague has been inspiring and pushing me to seek answers to the hard questions in neuroscience. Our regular conversations are a treasure. Monica Capra and Charles Noussair taught me how to think like an economist, while Michael Prietula and Tony Lemieux brought valuable insights about the effects of media and information on the brain and were instrumental to the design of several experiments. Scott Atran showed me how to approach questions of morality from a humble perspective. Philippe Rochat has been a source of enthusiasm and support during the writing of this book. He indulged my wild ideas while pursuing his own. We egged each other on to write during the Covid years.

None of the work I described would have been possible without grants. Individual program officers have made a huge difference in

257

what I have been able to study. I am especially grateful to the late Terry Lyons, who took me under his wing, introducing me to a side of the Department of Defense that I didn't know existed. Ivy Estabrooke was instrumental in funding several of our proposals, and so, too, was Bill Casebeer.

I am especially grateful to Michelle Tessler for finding a home for this book, before it was what it is now. Matching me with T. J. Kelleher at Basic Books was the best thing that happened for this project. This was the second book with T. J., and I didn't appreciate until this one how much trust is inherent in the author-editor relationship. I guess it takes a couple of books to get to that point. Tisse Takagi did an amazing job on the detailed editing. Somehow she made me sound more like me (or, at least, who I was).

Finally, I am grateful to my wife, Kathleen. She endured another bout of my complaints of this being the "last book." More importantly, she embraced our new adventures. Who knows what we'll become?

Notes

Introduction. The Self Delusion

1. David J. Chalmers, *The Conscious Mind: In Search of a Fundamental Theory* (New York: Oxford University Press, 1996).

2. Stephen King, *On Writing: A Memoir of the Craft* (New York: Scribner, 2000).

3. Untitled section written by "The Editor's Editor," *The Editor: The Journal of Information for Literary Workers* 45, no. 4 (February 24, 1917): 175, 176. Edited by William R. Kane, published by The Editor Company, Ridgewood, NJ.

4. H. Porter Abbott, *The Cambridge Introduction to Narrative*, 3rd ed. (Cambridge: Cambridge University Press, 2021).

Chapter 1. You Are a Simulation

1. Roger Brown and James Kulik, "Flashbulb Memories," *Cognition* 5, no. 1 (1977): 73–99.

2. William Hirst, Elizabeth A. Phelps, Robert Meksin, Chandan J. Vaidya, Marcia K. Johnson, Karen J. Mitchell, Randy L. Buckner, et al., "A Ten-Year Follow-Up of a Study of Memory for the Attack of September 11, 2001: Flashbulb Memories and Memories for Flashbulb Events," *Journal of Experimental Psychology: General* 144, no. 3 (2015): 604–623.

Chapter 2. Early Memories

1. Larry R. Squire and Stuart Zola-Morgan, "Memory: Brain Systems and Behavior," *Trends in Neurosciences* 11, no. 4 (1988): 170–175; Larry R. Squire, "Memory Systems of the Brain: A Brief History and Current Perspective," *Neurobiology of Learning and Memory* 82, no. 3 (2004): 171–177.

2. Michael D. Rugg, Jeffrey D. Johnson, Heekyeong Park, and Melina R. Uncapher, "Encoding-Retrieval Overlap in Human Episodic Memory: A Functional Neuroimaging Perspective," *Progress in Brain Research* 169 (2008): 339–352.

3. Catherine Lebel and Christian Beaulieu, "Longitudinal Development of Human Brain Wiring Continues from Childhood into Adulthood," *Journal of Neuroscience* 31, no. 30 (2011): 10937–10947; Jessica Dubois, Ghislaine Dehaene-Lambertz, Muriel Perrin, Jean-François Mangin, Yann Cointepas, Edouard Duchesnay, Denis Le Bihan, and Lucie Hertz-Pannier, "Asynchrony of the Early Maturation of White Matter Bundles in Healthy Infants: Quantitative Landmarks Revealed Noninvasively by Diffusion Tensor Imaging," *Human Brain Mapping* 29, no. 1 (2008): 14–27.

4. Robyn Fivush and Nina R. Hamond, "Autobiographical Memory Across the Preschool Years: Toward Reconceptualizing Childhood Amnesia," in *Knowing and Remembering in Young Children*, ed. Robyn Fivush and Judith A. Hudson (Cambridge: Cambridge University Press, 1990), 223–248.

5. JoNell A. Usher and Ulric Neisser, "Childhood Amnesia and the Beginnings of Memory for Four Early Life Events," *Journal of Experimental Psychology: General* 122, no. 2 (1993): 155–165.

6. Emily Sutcliffe Cleveland and Elaine Reese, "Children Remember Early Childhood: Long-Term Recall Across the Offset of Childhood Amnesia," *Applied Cognitive Psychology* 22, no. 1 (2008): 127–142; Elaine Reese, Fiona Jack, and Naomi White, "Origins of Adolescents' Autobiographical Memories," *Cognitive Development* 25, no. 4 (2010): 352–367.

7. Reese, Jack, and White, "Origins of Adolescents' Autobiographical Memories," 364.

8. Susan Engel, *The Stories Children Tell: Making Sense of the Narratives of Childhood* (New York: Henry Holt and Company, 1995).

9. Engel, *Stories Children Tell*, 92.

10. Ernest Hemingway, *The Sun Also Rises* (New York: Scribner, 1926), 198.

Chapter 3. Compression

1. If technology ever allows consciousness to be downloaded into a computer, it will be, at best, a lo-fi facsimile of the person. Maybe it will be better than the memories we hold of each other or the stories that have been told

through millennia, or maybe it will be just as compressed and artificial as a pixilated bitmap.

2. Frederic C. Bartlett, *Remembering: A Study in Experimental and Social Psychology* (Cambridge: Cambridge University Press, 1932).

3. Asaf Gilboa and Hannah Marlatte, "Neurobiology of Schemas and Schema-Mediated Memory," *Trends in Cognitive Sciences* 21, no. 8 (2017): 618–631.

4. Asaf Gilboa and Morris Moscovitch, "Ventromedial Prefrontal Cortex Generates Pre-Stimulus Theta Coherence Desynchronization: A Schema In-stantiation Hypothesis," *Cortex* 87 (2017): 16–30.

5. Olivier Jeunehomme and Arnaud D'Argembeau, "Event Segmentation and the Temporal Compression of Experience in Episodic Memory," *Psychological Research* 84, no. 2 (2020): 481–490.

Chapter 4. The Bayesian Brain

1. Daniel Kersten, Pascal Mamassian, and Alan Yuille, "Object Percep-tion as Bayesian Inference," *Annual Review of Psychology* 55 (2004): 271–304; C. Alejandro Parraga, Tom Troscianko, and David J. Tolhurst, "The Human Visual System Is Optimised for Processing the Spatial Information in Natural Visual Images," *Current Biology* 10, no. 1 (2000): 35–38.

2. Gergő Orbán, Pietro Berkes, József Fiser, and Máté Lengyel, "Neural Variability and Sampling-Based Probabilistic Representations in the Visual Cortex," *Neuron* 92, no. 2 (2016): 530–543; Robbe L. T. Goris, J. Anthony Movshon, and Eero P. Simoncelli, "Partitioning Neuronal Variability," *Nature Neuroscience* 17, no. 6 (2014): 858–865; A. Aldo Faisal, Luc P. J. Selen, and Daniel M. Wolpert, "Noise in the Nervous System," *Nature Reviews Neuroscience* 9, no. 4 (2008): 292–303.

3. Adam N. Sanborn and Nick Chater, "Bayesian Brains Without Proba-bilities," *Trends in Cognitive Sciences* 20, no. 12 (2016): 883–893.

4. Vincent Hayward, "A Brief Taxonomy of Tactile Illusions and Demon-strations That Can Be Done in a Hardware Store," *Brain Research Bulletin* 75, no. 6 (2008): 742–752.

5. Rebecca Boehme, Steven Hauser, Gregory J. Gerling, Markus Heilig, and Håkan Olausson, "Distinction of Self-Produced Touch and Social Touch at Cortical and Spinal Cord Levels," *Proceedings of the National Academy of Sciences* 116, no. 6 (2019): 2290–2299.

Chapter 5. The Bayesian Self

1. Franco Bertossa, Marco Besa, Roberto Ferrari, and Francesca Ferri, "Point Zero: A Phenomenological Inquiry into the Seat of Consciousness," *Perceptual and Motor Skills* 107, no. 2 (2008): 323–335; Jakub Limanowski and Heiko Hecht, "Where Do We Stand on Locating the Self?" *Psychology* 2,

no. 4 (2011): 312; Christina Starmans and Paul Bloom, "Windows to the Soul: Children and Adults See the Eyes as the Location of the Self," *Cognition* 123, no. 2 (2012): 313–318.

2. Lauri Nummenmaa, Riitta Hari, Jari K. Hietanen, and Enrico Glerean, "Maps of Subjective Feelings," *Proceedings of the National Academy of Sciences* 115, no. 37 (2018): 9198–9203.

3. Hadley Cantril and William A. Hunt, "Emotional Effects Produced by Injection of Adrenalin," *American Journal of Psychology* 44 (1932): 300–307.

4. Lisa Feldman Barrett, *How Emotions Are Made: The Secret Life of the Brain* (New York: Houghton Mifflin Harcourt, 2017).

5. Jaak Panksepp, *Affective Neuroscience: The Foundations of Human and Animal Emotions* (New York: Oxford University Press, 2004).

6. Jaak Panksepp, "The Basic Emotional Circuits of Mammalian Brains: Do Animals Have Affective Lives?" *Neuroscience & Biobehavioral Reviews* 35, no. 9 (2011): 1791–1804.

7. Charles Darwin, *The Expression of the Emotions in Man and Animals* (London: John Murray, 1872).

8. Ralph Adolphs, "How Should Neuroscience Study Emotions? By Distinguishing Emotion States, Concepts, and Experiences," *Social Cognitive and Affective Neuroscience* 12, no. 1 (2017): 24–31.

9. Shaun Gallagher, "Philosophical Conceptions of the Self: Implications for Cognitive Science," *Trends in Cognitive Sciences* 4, no. 1 (2000): 14–21; Jakob Hohwy, "The Sense of Self in the Phenomenology of Agency and Perception," *Psyche* 13, no. 1 (2007): 1–20.

10. Matthew Botvinick and Jonathan Cohen, "Rubber Hands 'Feel' Touch That Eyes See," *Nature* 391, no. 6669 (1998): 756.

11. Sandra Blakeslee and Matthew Blakeslee, *The Body Has a Mind of Its Own: How Body Maps in Your Brain Help You Do (Almost) Everything Better* (New York: Random House, 2007).

12. Daniel C. Dennett, *Consciousness Explained* (Boston: Little, Brown, 1991).

13. Georg Northoff, Alexander Heinzel, Moritz De Greck, Felix Bermpohl, Henrik Dobrowolny, and Jaak Panksepp, "Self-Referential Processing in Our Brain—a Meta-Analysis of Imaging Studies on the Self," *Neuroimage* 31, no. 1 (2006): 440–457; Jie Sui and Glyn W. Humphreys, "The Integrative Self: How Self-Reference Integrates Perception and Memory," *Trends in Cognitive Sciences* 19, no. 12 (2015): 719–728.

14. Marcus E. Raichle, "The Brain's Default Mode Network," *Annual Review of Neuroscience* 38 (2015): 433–447.

15. Michael D. Greicius, Vesa Kiviniemi, Osmo Tervonen, Vilho Vainionpää, Seppo Alahuhta, Allan L. Reiss, and Vinod Menon, "Persistent Default-Mode Network Connectivity During Light Sedation," *Human Brain Mapping* 29, no. 7 (2008): 839–847.

16. Mihaly Csikszentmihalyi, *Flow: The Psychology of Optimal Experience* (New York: Harper & Row, 1990).

17. Anil K. Seth and Karl J. Friston, "Active Interoceptive Inference and the Emotional Brain," *Philosophical Transactions of the Royal Society B: Biological Sciences* 371, no. 1708 (2016): 20160007.

18. Rosa Lafer-Sousa, Katherine L. Hermann, and Bevil R. Conway, "Striking Individual Differences in Color Perception Uncovered by 'the Dress' Photograph," *Current Biology* 25, no. 13 (2015): R545–R546.

Chapter 6. All the Yous

1. Morton Prince, *The Dissociation of a Personality. A Biographical Study in Abnormal Psychology* (New York: Longmans, Green, 1906).

2. Pierre Janet, *The Major Symptoms of Hysteria. Fifteen Lectures Given in the Medical School of Harvard University* (New York: Macmillan Company, 1907).

3. Henri Frédéric Ellenberger, "The Story of 'Anna O': A Critical Review with New Data," *Journal of the History of the Behavioral Sciences* (1972).

4. Ellenberger cites Jung, "Notes on the Seminar in Analytical Psychology Conducted by C. G. Jung" (unpublished typescript, Zurich, March 23–July 6, 1925). Arranged by Members of the Class, Zurich, 1926.

5. Ernest Jones, *The Life and Work of Sigmund Freud*, Vol. I (New York: Basic Books, 1953). The pseudonym was made by shifting the initials of the patient from B. P. to A. O.

6. Corbett H. Thigpen and Hervey Cleckley, "A Case of Multiple Personality," *Journal of Abnormal and Social Psychology* 49, no. 1 (1954): 135.

7. Debbie Nathan, *Sybil Exposed: The Extraordinary Story Behind the Famous Multiple Personality Case* (New York: Simon & Schuster, 2011).

8. Nathan, *Sybil Exposed*, 88.

9. Nathan, *Sybil Exposed*, Introduction.

10. Y. A. Aderibigbe, R. M. Bloch, and W. R. Walker, "Prevalence of Depersonalization and Derealization Experiences in a Rural Population," *Social Psychiatry and Psychiatric Epidemiology* 36, no. 2 (2001): 63–69.

11. Steven Jay Lynn, Scott O. Lilienfeld, Harald Merckelbach, Timo Giesbrecht, and Dalena van der Kloet, "Dissociation and Dissociative Disorders: Challenging Conventional Wisdom," *Current Directions in Psychological Science* 21, no. 1 (2012): 48–53.

Chapter 7. The Evolution of Narrative

1. Gregory Berns, *How Dogs Love Us: A Neuroscientist and His Adopted Dog Decode the Canine Brain* (New York: Houghton Mifflin Harcourt, 2013); Gregory Berns, *What It's Like to Be a Dog: And Other Adventures in Animal Neuroscience* (New York: Basic Books, 2017).

2. Stuart A. Vyse, *Believing in Magic: The Psychology of Superstition*, updated ed. (Oxford: Oxford University Press, 2013), 81.

3. J. L. Evenden and T. W. Robbins, "Win-Stay Behaviour in the Rat," *Quarterly Journal of Experimental Psychology Section B* 36, no. 1b (1984): 1–26.

4. Burrhus Frederic Skinner, "'Superstition' in the Pigeon," *Journal of Experimental Psychology* 38, no. 2 (1948): 168.

5. Gregory A. Wagner and Edward K. Morris, "'Superstitious' Behavior in Children," *Psychological Record* 37, no. 4 (1987): 471–488.

6. Koichi Ono, "Superstitious Behavior in Humans," *Journal of the Experimental Analysis of Behavior* 47, no. 3 (1987): 261–271.

7. Vyse, *Believing in Magic*, 135.

8. Events reconstructed from the following sources: "Nun's 1960 Recovery May Answer Prayers for Serra's Sainthood," by Mark I. Pinsky, *Los Angeles Times*, August 4, 1987; "Focus Now Shifts to Canonization of Serra," by Mark I. Pinsky, *Los Angeles Times*, October 1, 1988; "Catholic Down to the Bootstraps," by Anne Knight, Los Angeles Lay Catholic Mission, 1998.

9. Sue Ellen Wilcox, "Behind Every Saint Sister Boniface Meets with Pope John Paul II in Rome," *Chicago Heights Star*, November 12, 1987.

10. "Serra's Miracle Nun" (interview with Sister Boniface Dyrda), filmed by KGO-TV (ABC7), September 15, 1988, YouTube video, 4:30, www.youtube.com/watch?v=aRBSc7mL6cU (10/24/2019).

11. Francisco Palóu, *Life of Fray Junípero Serra*, Vol. 3 (Washington, DC: Academy of American Franciscan History, 1955).

12. M. I. Pinksy, "Focus Now Shifts to Canonization of Serra," *Los Angeles Times*, October 1, 1988.

Chapter 8. Narrative Forms

1. Vladimir Iakovlevich Propp, *Morphology of the Folktale*, trans. Laurence Scott. Vol. 9 (Austin: University of Texas Press, 1968).

2. Joseph Campbell, *The Hero with a Thousand Faces*, Vol. 17 (Novato, CA: New World Library, 2008). Because Propp's book wasn't translated into English until 1958, Campbell's work had been out for almost a decade before most people learned of Propp.

3. Gerald I. Davis, *Gilgamesh: The New Translation* (Bridgeport, CT: Insignia Publishing, 2014).

4. Andrew J. Reagan, Lewis Mitchell, Dilan Kiley, Christopher M. Danforth, and Peter Sheridan Dodds, "The Emotional Arcs of Stories Are Dominated by Six Basic Shapes," *EPJ Data Science* 5, no. 1 (2016): 1–12.

5. Rahav Gabay, Boaz Hameiri, Tammy Rubel-Lifschitz, and Arie Nadler, "The Tendency for Interpersonal Victimhood: The Personality Construct and Its Consequences," *Personality and Individual Differences* 165 (2020): 110134.

6. Jeffrey A. Bridge, Joel B. Greenhouse, Donna Ruch, Jack Stevens, John Ackerman, Arielle H. Sheftall, Lisa M. Horowitz, Kelly J. Kelleher, and John V. Campo, "Association Between the Release of Netflix's *13 Reasons Why* and Suicide Rates in the United States: An Interrupted Time Series Analysis," *Journal of the American Academy of Child & Adolescent Psychiatry* 59, no. 2 (2020): 236–243.

7. Daniel Romer, "Reanalysis of the Bridge et al. Study of Suicide Following Release of *13 Reasons Why*," *PLoS One* 15, no. 1 (2020): e0227545.

Chapter 9. Flavors of You

1. Jean-Jacques Rousseau, *A Discourse on Inequality* (London: Penguin, 1985). But also see: Adam Smith, *The Theory of Moral Sentiments*, Vol. 1 (London: J. Richardson, 1822).

2. David Hume, *A Treatise of Human Nature* (Mineola, NY: Dover Publications, 2003). Also: Brian Skyrms, *The Stag Hunt and the Evolution of Social Structure* (Cambridge: Cambridge University Press, 2004).

3. Daniel C. Dennett, "The Self as the Center of Narrative Gravity," in *Self and Consciousness: Multiple Perspectives*, ed. Frank S. Kessel, Pamela M. Cole, Dale L. Johnson, and Milton D. Hakel (Hillsdale, NJ: Lawrence Erlbaum, 1992), 103–115; Simon Baron-Cohen, "The Autistic Child's Theory of Mind: A Case of Specific Developmental Delay," *Journal of Child Psychology and Psychiatry* 30, no. 2 (1989): 285–297.

4. Helen L. Gallagher and Christopher D. Frith, "Functional Imaging of 'Theory of Mind,'" *Trends in Cognitive Sciences* 7, no. 2 (2003): 77–83; Sara M. Schaafsma, Donald W. Pfaff, Robert P. Spunt, and Ralph Adolphs, "Deconstructing and Reconstructing Theory of Mind," *Trends in Cognitive Sciences* 19, no. 2 (2015): 65–72.

5. W. Gavin Ekins, Ricardo Caceda, C. Monica Capra, and Gregory S. Berns, "You Cannot Gamble on Others: Dissociable Systems for Strategic Uncertainty and Risk in the Brain," *Journal of Economic Behavior & Organization* 94 (2013): 222–233.

6. Birgit A. Völlm, Alexander N. W. Taylor, Paul Richardson, Rhiannon Corcoran, John Stirling, Shane McKie, John F. W. Deakin, and Rebecca Elliott, "Neuronal Correlates of Theory of Mind and Empathy: A Functional Magnetic Resonance Imaging Study in a Nonverbal Task," *NeuroImage* 29, no. 1 (2006): 90–98.

Chapter 10. The Evolution of Groupthink

1. Solomon E. Asch, "Effects of Group Pressure Upon the Modification and Distortion of Judgments," in *Groups, Leadership and Men: Research in Human Relations*, ed. Harold Guetzkow (Pittsburgh, PA: Carnegie Press, 1951).

2. Gregory S. Berns, Jonathan Chappelow, Caroline F. Zink, Giuseppe Pagnoni, Megan E. Martin-Skurski, and Jim Richards, "Neurobiological Correlates of Social Conformity and Independence During Mental Rotation," *Biological Psychiatry* 58, no. 3 (2005): 245–253.

3. Stanley Milgram, "Behavioral Study of Obedience," *Journal of Abnormal and Social Psychology* 67, no. 4 (1963): 371–378.

4. The experiments were replicated in the modern era in Poland. Dariusz Doliński, Tomasz Grzyb, Michał Folwarczny, Patrycja Grzybała, Karolina Krzyszycha, Karolina Martynowska, and Jakub Trojanowski, "Would You Deliver an Electric Shock in 2015? Obedience in the Experimental Paradigm Developed by Stanley Milgram in the 50 Years Following the Original Studies," *Social Psychological and Personality Science* 8, no. 8 (2017): 927–933.

5. Daniel Kahneman and Amos Tversky, "Prospect Theory: An Analysis of Decision Under Risk," *Econometrica* 47, no. 2 (1979): 263–292.

6. Jan B. Engelmann, C. Monica Capra, Charles Noussair, and Gregory S. Berns, "Expert Financial Advice Neurobiologically 'Offloads' Financial Decision-Making Under Risk," *PLoS One* 4, no. 3 (2009): e4957.

7. The strategy is called "satisficing." Herbert A. Simon, "Rational Choice and the Structure of the Environment," *Psychological Review* 63, no. 2 (1956): 129–138.

8. James Surowiecki, *The Wisdom of Crowds: Why the Many Are Smarter Than the Few and How Collective Wisdom Shapes Business, Economies, Societies and Nations* (New York: Doubleday, 2004).

9. Francis Galton, "Vox Populi (the Wisdom of Crowds)," *Nature* 75, no. 7 (1907): 450–451.

10. Guido Biele, Jörg Rieskamp, Lea K. Krugel, and Hauke R. Heekeren, "The Neural Basis of Following Advice," *PLoS Biology* 9, no. 6 (2011): e1001089.

11. Morton Deutsch and Harold B. Gerard, "A Study of Normative and Informational Social Influences upon Individual Judgment," *Journal of Abnormal and Social Psychology* 51, no. 3 (1955): 629–636; Robert B. Cialdini and Noah J. Goldstein, "Social Influence: Compliance and Conformity," *Annual Review of Psychology* 55 (2004): 591–621.

12. Not all researchers agree with this logic. The weakness in my argument is its presumed dependence on identifying a specific cognitive process from the pattern of brain activity. Russ Poldrack, a professor at Stanford and one of the early pioneers in brain imaging, wrote an influential paper about the difficulty in deducing mental processes from brain activity (Russell A. Poldrack, "Can Cognitive Processes Be Inferred from Neuroimaging Data?" *Trends in Cognitive Sciences* 10, no. 2 [2006]: 59–63). He argued that because the brain is so interconnected, individual parts may have more than one function. So, what an individual region is doing at any one time depends not only on its own activity but also on what other connected regions are doing.

Because of this interconnectedness, Poldrack said one can't infer a mental process from activity in a single region by itself. He called the problem "reverse inference." Poldrack, though, created a potential solution by collecting thousands of fMRI experiments into a database called *Neurosynth*. We can now search the database by cognitive process to see which brain regions are associated with specific psychological terms. And we can search it in reverse inference mode to see the probability that a particular region is associated with a particular psychological term.

13. Gregory S. Berns, C. Monica Capra, Sara Moore, and Charles Noussair, "Neural Mechanisms of the Influence of Popularity on Adolescent Ratings of Music," *NeuroImage* 49, no. 3 (2010): 2687–2696.

14. Daniel K. Campbell-Meiklejohn, Dominik R. Bach, Andreas Roepstorff, Raymond J. Dolan, and Chris D. Frith, "How the Opinion of Others Affects Our Valuation of Objects," *Current Biology* 20, no. 13 (2010): 1165–1170.

15. Jamil Zaki, Jessica Schirmer, and Jason P Mitchell, "Social Influence Modulates the Neural Computation of Value," *Psychological Science* 22, no. 7 (2011): 894–900; Vasily Klucharev, Kaisa Hytönen, Mark Rijpkema, Ale Smidts, and Guillén Fernández, "Reinforcement Learning Signal Predicts Social Conformity," *Neuron* 61, no. 1 (2009): 140–151.

16. Hilke Plassmann, John O'Doherty, Baba Shiv, and Antonio Rangel, "Marketing Actions Can Modulate Neural Representations of Experienced Pleasantness," *Proceedings of the National Academy of Sciences* 105, no. 3 (2008): 1050–1054; Mirre Stallen, Nicholas Borg, and Brian Knutson, "Brain Activity Foreshadows Stock Price Dynamics," *Journal of Neuroscience* 41, no. 14 (2021): 3266–3274.

17. Technically, log(sales).

18. Sharad Goel, Jake M. Hofman, Sébastien Lahaie, David M. Pennock, and Duncan J. Watts, "Predicting Consumer Behavior with Web Search," *Proceedings of the National Academy of Sciences* 107, no. 41 (2010): 17486–17490.

Chapter 11. Moral Backbone

1. Moin Syed and Kate C. McLean, "Erikson's Theory of Psychosocial Development," in *The Sage Encyclopedia of Intellectual and Developmental Disorders*, ed. Ellen Braaten (Los Angeles: SAGE Publications, 2018), 577–581.

2. Jesse Graham and Jonathan Haidt, "Sacred Values and Evil Adversaries: A Moral Foundations Approach," in *The Social Psychology of Morality: Exploring the Causes of Good and Evil*, ed. Mario Mikulincer and Phillip R. Shaver (Washington, DC: American Psychological Association, 2012), 11–31.

3. William D. Casebeer, "Moral Cognition and Its Neural Constituents," *Nature Reviews Neuroscience* 4, no. 10 (2003): 840–846.

4. Jeremy Bentham, *The Principles of Morals and Legislation* (1780; repr., Amherst, NY: Prometheus Books, 1988); John Stuart Mill, *Utilitarianism*, 4th ed. (London: Longmans, Green, Reader, and Dyer, 1871).

5. Immanuel Kant, *Groundwork of the Metaphysics of Morals* (1785; repr., Toronto: Broadview Press, 2005).

6. Gordon M. Becker, Morris H. DeGroot, and Jacob Marschak, "Measuring Utility by a Single-Response Sequential Method," *Behavioral Science* 9, no. 3 (1964): 226–232.

7. The full list of statements can be found in the supplementary information of our paper: Gregory S. Berns, Emily Bell, C. Monica Capra, Michael J. Prietula, Sara Moore, Brittany Anderson, Jeremy Ginges, and Scott Atran, "The Price of Your Soul: Neural Evidence for the Non-Utilitarian Representation of Sacred Values," *Philosophical Transactions of the Royal Society B: Biological Sciences* 367, no. 1589 (2012): 754–762.

8. Jamil P. Bhanji, Jennifer S. Beer, and Silvia A. Bunge, "Taking a Gamble or Playing by the Rules: Dissociable Prefrontal Systems Implicated in Probabilistic Versus Deterministic Rule-Based Decisions," *NeuroImage* 49, no. 2 (2010): 1810–1819; Jonathan D. Wallis, Kathleen C. Anderson, and Earl K. Miller, "Single Neurons in Prefrontal Cortex Encode Abstract Rules," *Nature* 411, no. 6840 (2001): 953–956.

9. Liane Young, Joan Albert Camprodon, Marc Hauser, Alvaro Pascual-Leone, and Rebecca Saxe, "Disruption of the Right Temporoparietal Junction with Transcranial Magnetic Stimulation Reduces the Role of Beliefs in Moral Judgments," *Proceedings of the National Academy of Sciences* 107, no. 15 (2010): 6753–6758.

10. Melanie Pincus, Lisa LaViers, Michael J. Prietula, and Gregory Berns, "The Conforming Brain and Deontological Resolve," *PLoS One* 9, no. 8 (2014): e106061.

Chapter 12. The Banality of a Brain

1. "Assault or Homicide," Centers for Disease Control and Prevention, www.cdc.gov/nchs/fastats/homicide.htm, updated January 5, 2022, accessed January 6, 2022. This figure does not include killings by police.

2. Deborah W. Denno, "Revisiting the Legal Link Between Genetics and Crime," *Law and Contemporary Problems* 69, nos. 1/2 (2006): 209–257.

3. Han G. Brunner, M. Nelen, X. O. Breakefield, H. H. Ropers, and B. A. Van Oost, "Abnormal Behavior Associated with a Point Mutation in the Structural Gene for Monoamine Oxidase A," *Science* 262, no. 5133 (1993): 578–580.

4. Nigel Eastman and Colin Campbell, "Neuroscience and Legal Determination of Criminal Responsibility," *Nature Reviews Neuroscience* 7, no. 4 (2006): 311–318.

5. Charles Darwin, *The Expression of the Emotions in Man and Animals* (London: John Murray, 1872).

6. More recently, some researchers have questioned the universality of emotional expression. See: Lisa Feldman Barrett, *How Emotions Are Made: The Secret Life of the Brain* (New York: Houghton Mifflin Harcourt, 2017).

7. Paul Ekman, Wallace V. Friesen, Maureen O'Sullivan, Anthony Chan, Irene Diacoyanni-Tarlatzis, Karl Heider, Rainer Krause, et al., "Universals and Cultural Differences in the Judgments of Facial Expressions of Emotion," *Journal of Personality and Social Psychology* 53, no. 4 (1987): 712–717.

8. Nancy Kanwisher, Josh McDermott, and Marvin M. Chun, "The Fusiform Face Area: A Module in Human Extrastriate Cortex Specialized for Face Perception," *Journal of Neuroscience* 17, no. 11 (1997): 4302–4311.

9. Paul J. Whalen, Scott L. Rauch, Nancy L. Etcoff, Sean C. McInerney, Michael B. Lee, and Michael A. Jenike, "Masked Presentations of Emotional Facial Expressions Modulate Amygdala Activity Without Explicit Knowledge," *Journal of Neuroscience* 18, no. 1 (1998): 411–418.

10. Adrian Raine, Monte S. Buchsbaum, Jill Stanley, Steven Lottenberg, Leonard Abel, and Jacqueline Stoddard, "Selective Reductions in Prefrontal Glucose Metabolism in Murderers," *Biological Psychiatry* 36, no. 6 (1994): 365–373.

11. Peter F. Liddle, Kent A. Kiehl, and Andra M. Smith, "Event-Related fMRI Study of Response Inhibition," *Human Brain Mapping* 12, no. 2 (2001): 100–109.

12. Eyal Aharoni, Gina M. Vincent, Carla L. Harenski, Vince D. Calhoun, Walter Sinnott-Armstrong, Michael S. Gazzaniga, and Kent A. Kiehl, "Neuroprediction of Future Rearrest," *Proceedings of the National Academy of Sciences* 110, no. 15 (2013): 6223–6228.

13. Hannah Arendt, *Eichmann in Jerusalem: A Report on the Banality of Evil* (New York: Viking Press, 1963).

Chapter 13. The Man with Half a Brain

1. Karl Jaspers, *General Psychopathology*, trans. J. Hoenig and Marian W. Hamilton (Manchester, UK: Manchester University Press, 1963); Hugh Jones, Philippe Delespaul, and Jim van Os, "Jaspers Was Right After All—Delusions Are Distinct from Normal Beliefs," *British Journal of Psychiatry* 183, no. 4 (2003): 285–286.

2. Herbert Y. Meltzer and Stephen M. Stahl, "The Dopamine Hypothesis of Schizophrenia: A Review," *Schizophrenia Bulletin* 2, no. 1 (1976): 19–76; Philip Seeman, "Dopamine Receptors and the Dopamine Hypothesis of Schizophrenia," *Synapse* 1, no. 2 (1987): 133–152; Stephen M. Stahl, "Beyond the Dopamine Hypothesis of Schizophrenia to Three Neural Networks of Psychosis: Dopamine, Serotonin, and Glutamate," *CNS Spectrums* 23, no. 3 (2018): 187–191.

3. Heavensgate.com.

4. Kenneth S. Kendler, Timothy J. Gallagher, Jamie M. Abelson, and Ronald C. Kessler, "Lifetime Prevalence, Demographic Risk Factors, and Diagnostic Validity of Nonaffective Psychosis as Assessed in a US Community Sample: The National Comorbidity Survey," *Archives of General Psychiatry* 53, no. 11 (1996): 1022–1031.

5. Mark Shevlin, Jamie Murphy, Martin J. Dorahy, and Gary Adamson, "The Distribution of Positive Psychosis-Like Symptoms in the Population: A Latent Class Analysis of the National Comorbidity Survey," *Schizophrenia Research* 89, nos. 1–3 (2007): 101–109.

6. Emmanuelle Peters, Samantha Day, Jacqueline McKenna, and Gilli Orbach, "Delusional Ideation in Religious and Psychotic Populations," *British Journal of Clinical Psychology* 38, no. 1 (1999): 83–96.

7. Louise C. Johns and Jim van Os, "The Continuity of Psychotic Experiences in the General Population," *Clinical Psychology Review* 21, no. 8 (2001): 1125–1141.

8. I explored this in depth in Gregory Berns, *Iconoclast: A Neuroscientist Reveals How to Think Differently* (Cambridge, MA: Harvard Business Press, 2008).

Chapter 14. Novels on the Brain

1. Stephen King, Introduction, in *Lord of the Flies*, Centenary ed. (London: Penguin, 2011).

2. Marcus E. Raichle, "The Brain's Default Mode Network," *Annual Review of Neuroscience* 38 (2015): 433–447.

3. Jessica R. Andrews-Hanna, Jay S. Reidler, Christine Huang, and Randy L. Buckner, "Evidence for the Default Network's Role in Spontaneous Cognition," *Journal of Neurophysiology* 104, no. 1 (2010): 322–335; Kalina Christoff, Alan M. Gordon, Jonathan Smallwood, Rachelle Smith, and Jonathan W. Schooler, "Experience Sampling During fMRI Reveals Default Network and Executive System Contributions to Mind Wandering," *Proceedings of the National Academy of Sciences* 106, no. 21 (2009): 8719–8724.

4. Michael D. Greicius, Vesa Kiviniemi, Osmo Tervonen, Vilho Vainionpää, Seppo Alahuhta, Allan L. Reiss, and Vinod Menon, "Persistent Default-Mode Network Connectivity During Light Sedation," *Human Brain Mapping* 29, no. 7 (2008): 839–847.

5. Uri Hasson, Howard C. Nusbaum, and Steven L. Small, "Task-Dependent Organization of Brain Regions Active During Rest," *Proceedings of the National Academy of Sciences* 106, no. 26 (2009): 10841–10846.

6. Allyson P. Mackey, Alison T. Miller Singley, and Silvia A. Bunge, "Intensive Reasoning Training Alters Patterns of Brain Connectivity at Rest," *Journal of Neuroscience* 33, no. 11 (2013): 4796–4803.

7. Gregory S. Berns, Kristina Blaine, Michael J. Prietula, and Brandon E. Pye, "Short- and Long-Term Effects of a Novel on Connectivity in the Brain," *Brain Connectivity* 3, no. 6 (2013): 590–600.

8. Robert Harris, *Pompeii: A Novel* (New York: Random House, 2003).

9. Gustavo Deco, Viktor K. Jirsa, and Anthony R. McIntosh, "Emerging Concepts for the Dynamical Organization of Resting-State Activity in the Brain," *Nature Reviews Neuroscience* 12, no. 1 (2011): 43–56.

10. Lisa Aziz-Zadeh and Antonio Damasio, "Embodied Semantics for Actions: Findings from Functional Brain Imaging," *Journal of Physiology-Paris* 102, nos. 1–3 (2008): 35–39.

11. Simon Lacey, Randall Stilla, and Krish Sathian, "Metaphorically Feeling: Comprehending Textural Metaphors Activates Somatosensory Cortex," *Brain and Language* 120, no. 3 (2012): 416–421.

12. Robert Stickgold, "Sleep-Dependent Memory Consolidation," *Nature* 437, no. 7063 (2005): 1272–1278.

13. The stereotype of a teenage boy, playing games in his room into the early morning hours, is not true. In 2019, 65 percent of American adults played video games, and the average age was thirty-three ("2019 Essential Facts About the Computer and Video Game Industry," Entertainment Software Association, 2019, www.theesa.com/resource/essential-facts-about-the-computer-and-video-game-industry-2019/). Fifty-four percent were male and 46 percent were female. Video games are also a source of social connection, with 63 percent playing with others. Millennial males preferred shooter and sports games, like *Call of Duty* and *Fortnite* and *Madden NFL*. Millennial females preferred casual and action games, like *Candy Crush* and *Tomb Raider*. Boomers of both genders tended to play puzzle games, like *Solitaire* and *Scrabble*.

14. Wei Pan, Xuemei Gao, Shuo Shi, Fuqu Liu, and Chao Li, "Spontaneous Brain Activity Did Not Show the Effect of Violent Video Games on Aggression: A Resting-State fMRI Study," *Frontiers in Psychology* 8 (2018): 2219.

Chapter 15. Hijacked Narratives

1. Jessica McBride, "'Plandemic' Movie: Fact-Checking the New COVID-19 Video," Heavy, May 8, 2020, https://heavy.com/news/2020/05/plandemic-video-fact-checking-true-false/.

2. "Mikki Willis Biography," IMDb, www.imdb.com/name/nm0932413/bio?ref_=nm_ov_bio_sm.

3. Martin Enserink and Jon Cohen, "Fact-Checking Judy Mikovits, the Controversial Virologist Attacking Anthony Fauci in a Viral Conspiracy Video," *Science* 8 (2020).

4. Josh Rottenberg and Stacy Perman, "Meet the Ojai Dad Who Made the Most Notorious Piece of Coronavirus Disinformation Yet," *Los Angeles Times*,

May 13, 2020, www.latimes.com/entertainment-arts/movies/story/2020-05 -13/plandemic-coronavirus-documentary-director-mikki-willis-mikovits.

5. Jim Carrey (@JimCarrey), "California Gov says yes to poisoning more children . . .," Twitter, June 30, 2015, 7:03 p.m., https://twitter.com/JimCarrey /status/616049450243338240.

6. Josh Hafenbrack, "Trump: Autism Linked to Child Vaccinations," *South Florida Sun Sentinel*, December 28, 2007, www.sun-sentinel.com/sfl -mtblog-2007-12-trump_autism_linked_to_child_v-story.html.

7. "Data & Statistics on Autism Spectrum Disorder," Centers for Disease Control and Prevention, updated December 21, 2021, www.cdc.gov /ncbddd/autism/data.html.

8. Nadja Durbach, "'They Might as Well Brand Us': Working-Class Resistance to Compulsory Vaccination in Victorian England," *Social History of Medicine* 13, no. 1 (2000): 45–63, www.historyofvaccines.org/index.php /content/articles/history-anti-vaccination-movements.

9. Robert M. Wolfe and Lisa K. Sharp, "Anti-Vaccinationists Past and Present," *BMJ* 325, no. 7361 (2002): 430–432.

10. Jeffrey P. Baker, "The Pertussis Vaccine Controversy in Great Britain, 1974–1986," *Vaccine* 21, nos. 25–26 (2003): 4003–4010.

11. Anna Kata, "A Postmodern Pandora's Box: Anti-Vaccination Misinformation on the Internet," *Vaccine* 28, no. 7 (2010): 1709–1716.

12. Karen M. Douglas, Robbie M. Sutton, and Aleksandra Cichocka, "The Psychology of Conspiracy Theories," *Current Directions in Psychological Science* 26, no. 6 (2017): 538–542.

13. Russell Hardin, "The Crippled Epistemology of Extremism," in *Political Extremism and Rationality*, ed. Albert Breton, Gianluigi Galeotti, Pierre Salmon, and Ronald Wintrobe (Cambridge: Cambridge University Press, 2002), 3–22.

14. Cornelius Adrian Vermeule and Cass Robert Sunstein, "Conspiracy Theories: Causes and Cures," *Journal of Political Philosophy* (2009): 202–227.

15. Chip Heath, Chris Bell, and Emily Sternberg, "Emotional Selection in Memes: The Case of Urban Legends," *Journal of Personality and Social Psychology* 81, no. 6 (2001): 1028.

16. Aleksandra Cichocka, Marta Marchlewska, and Agnieszka Golec de Zavala, "Does Self-Love or Self-Hate Predict Conspiracy Beliefs? Narcissism, Self-Esteem, and the Endorsement of Conspiracy Theories," *Social Psychological and Personality Science* 7, no. 2 (2016): 157–166.

17. Vincent C. Lombardi, Francis W. Ruscetti, Jaydip Das Gupta, Max A. Pfost, Kathryn S. Hagen, Daniel L. Peterson, Sandra K. Ruscetti, et al., "Detection of an Infectious Retrovirus, XMRV, in Blood Cells of Patients with Chronic Fatigue Syndrome," *Science* 326, no. 5952 (2009): 585–589; Graham Simmons, Simone A. Glynn, Anthony L. Komaroff, Judy A. Mikovits, Leslie H. Tobler, John Hackett, Ning Tang, et al., "Failure to

Confirm XMRV/MLVs in the Blood of Patients with Chronic Fatigue Syndrome: A Multi-Laboratory Study," *Science* 334, no. 6057 (2011): 814–817; Robert H. Silverman, Jaydip Das Gupta, Vincent C. Lombardi, Francis W. Ruscetti, Max A. Pfost, Kathryn S. Hagen, Daniel L. Peterson, et al., "Partial Retraction," *Science* 334, no. 6053 (2011): 176.

18. Jon Cohen and Martin Enserink, "False Positive," *Science* 333 (2011): 1694–1701.

19. Jocelyn J. Bélanger, Julie Caouette, Keren Sharvit, and Michelle Dugas, "The Psychology of Martyrdom: Making the Ultimate Sacrifice in the Name of a Cause," *Journal of Personality and Social Psychology* 107, no. 3 (2014): 494–515.

Chapter 16. No Regrets

1. Ruth M. J. Byrne, "Counterfactual Thought," *Annual Review of Psychology* 67 (2016): 135–157.

2. Graham Loomes and Robert Sugden, "Regret Theory: An Alternative Theory of Rational Choice Under Uncertainty," *Economic Journal* 92, no. 368 (1982): 805–824; Barbara Mellers, Alan Schwartz, and Ilana Ritov, "Emotion-Based Choice," *Journal of Experimental Psychology: General* 128, no. 3 (1999): 332–345.

3. Nathalie Camille, Giorgio Coricelli, Jerome Sallet, Pascale Pradat-Diehl, Jean-René Duhamel, and Angela Sirigu, "The Involvement of the Orbitofrontal Cortex in the Experience of Regret," *Science* 304, no. 5674 (2004): 1167–1170.

4. Giorgio Coricelli, Hugo D. Critchley, Mateus Joffily, John P. O'Doherty, Angela Sirigu, and Raymond J. Dolan, "Regret and Its Avoidance: A Neuroimaging Study of Choice Behavior," *Nature Neuroscience* 8, no. 9 (2005): 1255–1262.

5. Terry Lohrenz, Kevin McCabe, Colin F. Camerer, and P. Read Montague, "Neural Signature of Fictive Learning Signals in a Sequential Investment Task," *Proceedings of the National Academy of Sciences* 104, no. 22 (2007): 9493–9498.

6. Adam P. Steiner and A. David Redish, "Behavioral and Neurophysiological Correlates of Regret in Rat Decision-Making on a Neuroeconomic Task," *Nature Neuroscience* 17, no. 7 (2014): 995–1002.

7. For more details, see Gregory Berns, *What It's Like to Be a Dog: And Other Adventures in Animal Neuroscience* (New York: Basic Books, 2017).

8. Martin Zinkevich, Michael Johanson, Michael Bowling, and Carmelo Piccione, "Regret Minimization in Games with Incomplete Information," *Advances in Neural Information Processing Systems* 20 (2007): 1729–1736.

9. Michael Bowling, Neil Burch, Michael Johanson, and Oskari Tammelin, "Heads-Up Limit Hold'em Poker Is Solved," *Science* 347, no. 6218 (2015): 145–149.

10. Thomas Gilovich and Victoria Husted Medvec, "The Experience of Regret: What, When, and Why," *Psychological Review* 102, no. 2 (1995): 379–395.

11. Keith D. Markman, Matthew N. McMullen, and Ronald A. Elizaga, "Counterfactual Thinking, Persistence, and Performance: A Test of the Reflection and Evaluation Model," *Journal of Experimental Social Psychology* 44, no. 2 (2008): 421–428.

12. Byrne, "Counterfactual Thought," 135–157.

Chapter 17. The Good Life

1. Robert McKee, *Story: Substance, Structure, Style, and the Principles of Screenwriting* (New York: HarperCollins, 1997).

2. Jiyuan Yu, *The Ethics of Confucius and Aristotle: Mirrors of Virtue* (New York: Routledge, 2007).

3. Viktor E. Frankl, *Man's Search for Meaning: An Introduction to Logotherapy*, 3rd ed. (Boston: Beacon Press, 2016).

4. Lisa Cron, *Story Genius: How to Use Brain Science to Go Beyond Outlining and Write a Riveting Novel* (Berkeley, CA: Ten Speed Press, 2016).

5. Holistic Management International, http://holisticmanagement.org.

Chapter 18. The Future of You

1. Daniel L. Schacter, Roland G. Benoit, and Karl K. Szpunar, "Episodic Future Thinking: Mechanisms and Functions," *Current Opinion in Behavioral Sciences* 17 (2017): 41–50.

2. Daniel L. Schacter, Donna Rose Addis, and Randy L. Buckner, "Remembering the Past to Imagine the Future: The Prospective Brain," *Nature Reviews Neuroscience* 8, no. 9 (2007): 657–661; Peter Zeidman and Eleanor A. Maguire, "Anterior Hippocampus: The Anatomy of Perception, Imagination and Episodic Memory," *Nature Reviews Neuroscience* 17, no. 3 (2016): 173–182.

3. Kevin P. Madore, Brendan Gaesser, and Daniel L. Schacter, "Constructive Episodic Simulation: Dissociable Effects of a Specificity Induction on Remembering, Imagining, and Describing in Young and Older Adults," *Journal of Experimental Psychology: Learning, Memory, and Cognition* 40, no. 3 (2014): 609–622.

4. Emily Frith, Seungho Ryu, Minsoo Kang, and Paul D. Loprinzi, "Systematic Review of the Proposed Associations Between Physical Exercise and Creative Thinking," *Europe's Journal of Psychology* 15, no. 4 (2019): 858–877.

5. Michael Pollan, *How to Change Your Mind: What the New Science of Psychedelics Teaches Us About Consciousness, Dying, Addiction, Depression, and Transcendence* (New York: Penguin, 2018).

Index

GREGORY BERNS is a professor of psychology at Emory University, where he directs the Center for Neuropolicy and Facility for Education & Research in Neuroscience. He is the author of several books, including the *New York Times* and *Wall Street Journal* best seller *How Dogs Love Us*. He lives on a farm near Atlanta, Georgia.